Self-Assembled Molecules – New Kind of Protein Ligands

Irena Roterman • Leszek Konieczny

Editors

Self-Assembled Molecules – New Kind of Protein Ligands

Supramolecular Ligands

Editors
Irena Roterman
Department of Bioinformatics and
 Telemedicine
Jagiellonian University – Medical College
Krakow, Poland

Leszek Konieczny
Chair of Medical Biochemistry
Jagiellonian University – Medical College
Krakow, Poland

ISBN 978-3-319-65638-0   ISBN 978-3-319-65639-7   (eBook)
https://doi.org/10.1007/978-3-319-65639-7

Library of Congress Control Number: 2017953533

Printed on acid-free paper

This Springer imprint is published by Springer Nature
The registered company is Springer International Publishing AG
The registered company address is: Gewerbestrasse 11, 6330 Cham, Switzerland

Foreword

This collection of publications serves as the introduction to a new approach in biology and pharmacology: exploiting the peculiar properties of supramolecular systems, particularly ribbonlike micellar structures which constitute an entirely new category of protein ligands. The novelty of the problem is reflected by the specific character of such ligands but also by the way in which they bind to proteins – a mechanism unlike "classic" ligand binding. Among described problems of importance are enhancement of immune complexation by supramolecular ligands and their possible use as carriers for drugs. Many of those supramolecular compounds, including Congo red and Evans blue, have long been used as dyes and amyloid markers; however, we are only now beginning to understand their specific chemistry and interaction with proteins.

The micellar structure of supramolecular ligands enables intercalation of foreign particles, including drugs. This phenomenon is particularly interesting given the ligands' known affinity for antibodies – but only those engaged in immune complexes. Another important advantage is the strengthening of antigen-antibody interactions brought about by complexation of a supramolecular ligand. It therefore seems likely that supramolecular ligands will find use in immunotargeting.

Intercalation of customized complexones enables supramolecular ligands to inject metal ions into proteins – in order to provide contrast for EM imaging but also for therapeutic purposes.

The analysis of the complexation behavior of supramolecular ligands casts a new light on the phenomenon of amyloidogenesis. We can expect that further research into supramolecular systems will lead to a wider range of practical applications. The authors are predominantly biochemists involved in supramolecular compound application in biology and medicine. The ideas and results presented shall be of interest for researchers looking for new materials and methods in antibacterial therapy.

Krakow, Poland   Leszek Konieczny
Krakow, Poland   Irena Roterman

Acknowledgements

Chapters 1 and 5 Work financially supported by Collegium Medicum – Jagiellonian University grant system – grant # K/ZDS/006363.

Chapters 2, 3, 4, 6, and 7 We acknowledge the financial support from the National Science Centre, Poland (grant no. 2016/21/D/NZ1/02763) and from the project Interdisciplinary PhD Studies "Molecular sciences for medicine" (co-financed by the European Social Fund within the Human Capital Operational Programme) and Ministry of Science and Higher Education (grant no. K/DSC/001370).

Contents

Abbreviations

AC	Alizarin complex
AFM	Atomic force microscopy
B-J proteins	Bence-Jones proteins
CDR	Complementarity-determining regions
CR	Congo red
CYAB	Cetyltrimethylammonium bromide
DDAB	Dimethyldioctadecyl-ammonium bromide
DLS	Dynamic light scattering
DMSO	Dimethyl sulfoxide
DOX	Doxorubicin
DY	Direct yellow
DY28	Direct yellow 28
DY9	Direct yellow 9
EB	Evans blue
EDS	Energy dispersive spectroscopy
EM	Electron microscopy
EUCAST	European Committee on Antimicrobial Susceptibility Testing
FOD	Fuzzy oil drop model
Hb	Hemoglobin
ILs	Ionic liquids
IP	Propidium iodide
PA	*Pseudomonas aeruginosa*
PTFE membrane	Name of product
MDR	Multidrug resistant
MHA	Mueller-Hinton agar
MRSA	Methicillin-resistant *Staphylococcus aureus*
PDB	Protein Data Bank
RB	Rhodamine B
Rdf	Radial distribution functions
RILs	Room-temperature ionic liquids

RPMI 1640 medium	Name of compound
SDBC	Sodium dodecylbenzenesulfonate
SDS	Sodium dodecyl sulfate
SEM	Scanning electron microscopy
SRBC	Sheep red blood cells
SWNT	Single wall carbon nanotubes
TB	Trypan blue
TEM	Transmission electron microscopy, TEM
TY	Titan yellow
U937	Human lymphoid cell line

Chapter 1
Supramolecular Systems as Protein Ligands

Joanna Rybarska, Barbara Piekarska, Barbara Stopa, Grzegorz Zemanek, Leszek Konieczny, and Irena Roterman

Abstract The standard substrate complexation mechanism engages natural binding sites. In contrast, supramolecular structures may form complexes with proteins by penetrating in regions which are either naturally unstable or become temporarily accessible due to structural rearrangements related to the protein's function. This may result in enhancement of irreversible processes (e.g. immune complexation or complement activation) or inhibition of reversible processes (e.g. enzymatic catalysis). Only ribbon-like supramolecular structures may form complexes with proteins. Having anchored itself inside the protein, the supramolecular ligand is protected against environmental factors such as changes in pH. This type of interaction represents a unique, nonstandard phenomenon in the context of proteomics.

Keywords Protein dynamics and Congo red binding • Ribbon-like supramolecular micelles • Congo red as supramolecular dye • Self-assembled molecules form a unit protein ligand • Unity of self-assembled molecules • Congo red penetration to protein interior • Congo red complexation properties • Protection of bound Congo red by proteins

1.1 Mechanism of Complexation

Biological function is a critical aspect in proteomics, and is often defined as the capability to interact with specific ligands and form complexes. Protein ligands tend to be either small molecules or small fragments of larger systems. They bind to the target protein in a specific area called the active site (or active group). Typically, the

J. Rybarska (✉) • B. Piekarska • B. Stopa • G. Zemanek • L. Konieczny
Chair of Medical Biochemistry, Jagiellonian University – Medical College,
Kopernika 7, 31-034 Krakow, Poland
e-mail: mbstylin@cyf-kr.edu.pl; mbpiekar@cyf-kr.edu.pl; barbara.stopa@uj.edu.pl; grzegorz.
zemanek@uj.edu.pl; mbkoniec@cyf-kr.edu.pl

I. Roterman
Department of Bioinformatics and Telemedicine, Jagiellonian University – Medical College,
Łazarza 16, 31-530, Krakow, Poland
e-mail: myroterm@cyf-kr.edu.pl

© The Author(s) 2018
I. Roterman, L. Konieczny (eds.), *Self-Assembled Molecules – New Kind
of Protein Ligands*, https://doi.org/10.1007/978-3-319-65639-7_1

active site is a pocket where the ligand may directly contact the nonpolar interior of the protein – an environment which excludes water. The result is a stable complex and the ability to carry out reactions which would not be possible in an aqueous solution.

Proteins are generally incapable of interaction in areas other than their active sites, since tight packing of polypeptide chains prevents penetration of random ligands. Nevertheless, the protein is not a monolith: its dynamic nature means that under certain conditions the packing of polypeptide chains may undergo relaxation, enabling small molecules to penetrate protein interior [1–6]. Those ligands cannot form stable bonds due to low binding energy in an area otherwise unprepared for specific interaction with such compounds – to put it simply, a rigid molecule is not likely to exhibit good alignment with the conformation of a folded polypeptide chain. The high mobility of small ligands also discourages strong interactions.

In spite of the above, some supramolecular associations of organic compounds are able to penetrate and anchor themselves inside proteins. This unique property emerges as a result of association (or self-association) of individual molecules [7–12], and is linked to the flexible structure and large interaction surfaces exposed by supramolecular ligands.

The presence of noncovalent bonds in supramolecular structures allows their components to shift with respect to one another, resulting in an adaptive ligand which has greater conformational alignment capabilities than polymers or small organic molecules.

Ongoing progress in supramolecular chemistry opens new research avenues and highlights new uses for associative structures [13–19]. Currently, research effort focuses primarily on technological improvements, including novel sieves, adsorption systems or tools which exploit various mechanical effects. The goal of such initiatives is to synthesize suitable monomers (or polymeric structures), which can then associate with one another according to a predefined blueprint, producing complex supramolecular units (Fig. 1.1). Relatively little work has been done in the area of identifying biological applications of such structures.

Noncovalent association as a means of generating complex structures is a ubiquitous phenomenon in nature. One classic example is the formation of molecular membranes, where a counterbalance of positive and negative charges in the polar component of each monomer eliminates electrostatic repulsion and allows molecules to align side by side in water, forming sheets. Another example involves microtubules which consist of self-associating proteins (Fig. 1.2).

Not all associative organic structures are capable of attaching to proteins. In fact, only one specific type of supramolecular structure can form complexes with sufficient stability to contemplate practical applications: systems which adopt ribbon-like micellar conformations (Fig. 1.3).

Ribbon-like micelles are the result of association of flat, polyaromatic, elongated, symmetrical molecules with polar groups at either end. Examples include CR and EB [20, 21]. Ligands consisting of several such molecules may penetrate into the protein interior by exploiting local instabilities or gaps created through accidental displacement of polypeptide loops. The ligand typically anchors itself between

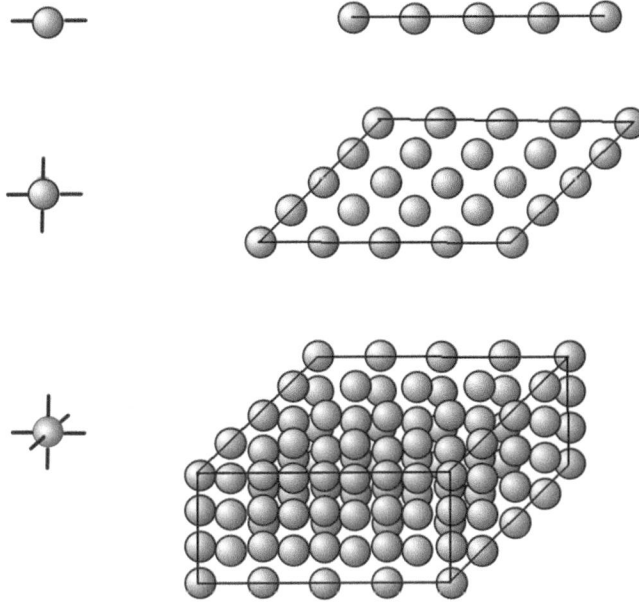

Fig. 1.1 Formation of one-, two- and three-dimensional associations of monomers – schematic presentation

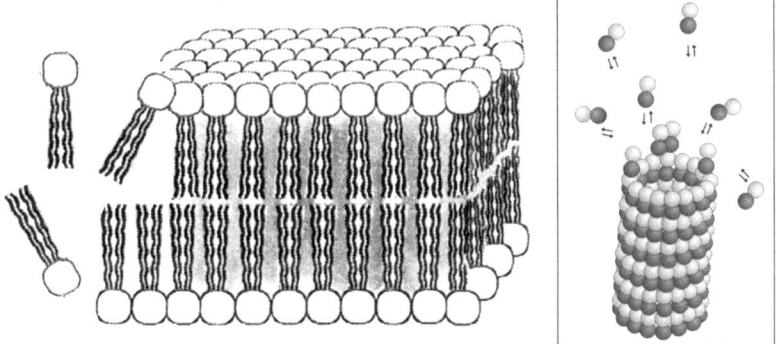

Fig. 1.2 Schematic depiction of cell membrane. Fencepost-like arrangement of phospholipid molecules enabled to close contact owing to charge neutrality. *Inset*: directed association of protein molecules – formation of microtubules

beta folds or random coils, since these two structural forms of polypeptide chain expose suitably large contact areas. Figure 1.4 illustrates the complexation process.

Owing to its structural flexibility, the supramolecular ligand may interact with proteins as the specific component – although its presence may also alter the target protein due to the large interaction area and strength of binding. Both structures adapt finally to each other, producing a stable bond [22].

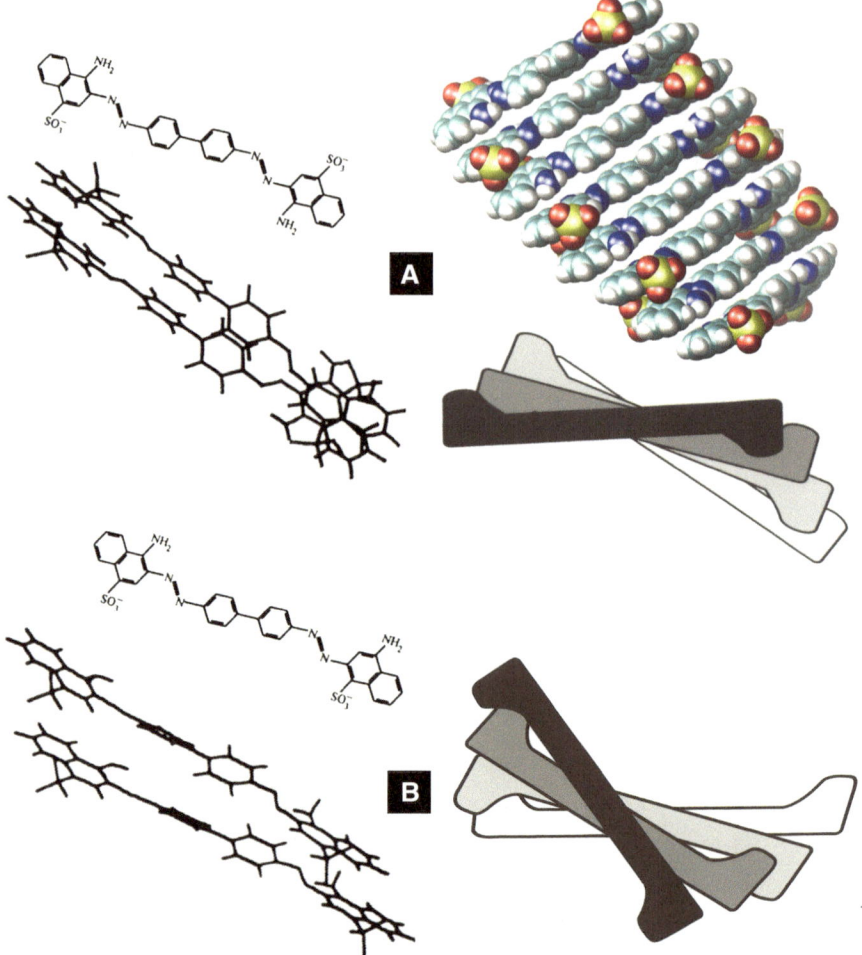

Fig. 1.3 Formation of supramolecular ribbon-like CR micelle. **A** - trans form of CR, **B** - cis form of CR

The large volume of supramolecular ligands undoubtedly hampers penetration. Consequently, supramolecular ligands prefer interaction with inherently unstable proteins – such as partly unfolded proteins and amyloids [23–34]. In some cases, however, even a tightly packed protein may – when binding its natural target ligand – undergo sufficient structural rearrangement to permit penetration of additional large supramolecular ligands penetrating outside of the primary binding site. This type of interaction, while temporary, often drastically modifies the function of the protein [35, 36].

Asymmetrical bipolar molecules which form supramolecular systems in water, such as detergents, form also complexes with proteins by penetrating into their hydrophobic areas; however, this mode of interaction differs from the one used by symmetrical molecules. With detergents penetration is diffuse and occurs wherever low polarity is present. This process produces major changes in the protein's sec-

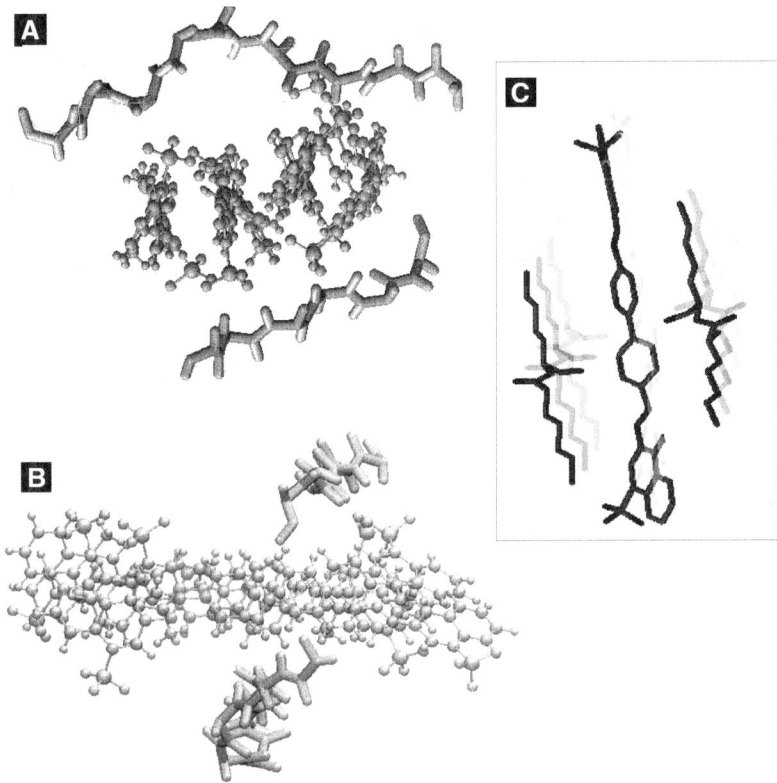

Fig. 1.4 CR/polypeptide complexation principle. (**A** and **B**) Molecular model (90-degree rotation). (**C**) Schematic view of CR (supramolecular) in complex with polylysine

ondary conformation, ultimately leading to denaturation. The unfolded skeleton of the protein is then reused by the supramolecular ligand as a seed for micellar aggregation. A typical example is the modification of polypeptide chains produced by SDS, commonly applied in polyacrylamide gel electrophoresis (Fig. 1.5) [37].

In contrast, flat, ribbon-like supramolecular ligands interact with proteins by "wedging" and produce no major changes in the protein's distribution of hydrophobicity. Despite forming a complex with the ribbon-like ligand, the protein retains its native interaction capabilities. This effect is reinforced by the stability and cohesive nature of the ligand itself (caused by strong intermolecular association). If the target protein acquires the ability to bind a supramolecular structure as a result of interacting with its natural ligand, then the presence of the supramolecular structure tends to stabilize the original protein/ligand complex. This occurs in the case of nonreversible interactions, such as between antigens and antibodies. On the other hand, supramolecular ligands are also able to inhibit the activity of enzymes by "freezing" them in their complex with the substrate. Such uncompetitive inhibition differs by a mechanism from that known as noncompetitive one. It indicates that ribbon-like supramolecular structures may also be of use in pharmacology as distinct inhibitors [38, 39] (Fig. 1.6).

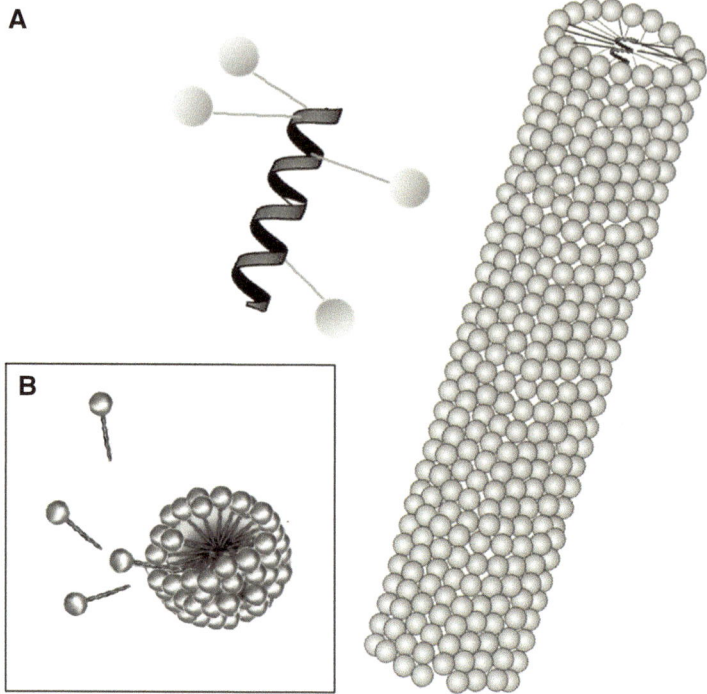

Fig. 1.5 Formation of rod-like structures of SDS: **A** - with protein backbone, **B** - without protein backbone

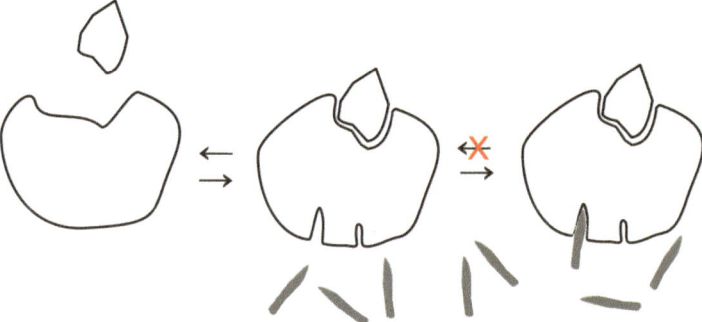

Fig. 1.6 Model view of enzyme inhibited by a supramolecular ligand (uncompetitive inhibition)

Uncompetitive inhibition is rarely encountered in nature as most inhibition processes are either competitive or noncompetitive in nature. In this specific case the inhibitor does not attach itself to the enzyme but rather to the entire enzyme-substrate complex, stabilizing it and negating the reversibility of complexation (Fig. 1.7).

It appears however that in order to form stable complexes with proteins without degrading their structure, supramolecular ligands must exhibit a ribbon-like conformation. It should also be noted that the distribution of polarity in a ribbon-like

Fig. 1.7 Mathematical
formulation of
uncompetitive inhibition

V_{max} · Maximal rate

K_m - Michaelis constat

K_i - Dissotiation constat

$[S]$ - Substrate concentration

$[I]$ - Inhibitor concentration

$$v = \frac{V_{max}\,[S]}{K_m + [S](1 + [I]/K_i)}$$

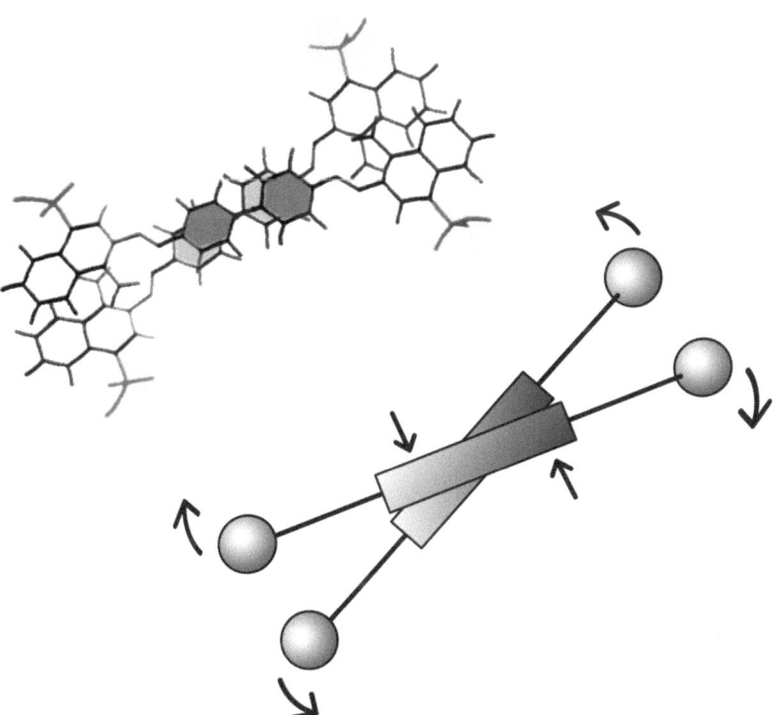

Fig. 1.8 Elongated association area of symmetric self-assembling molecules, necessary for formation of ribbon-like supramolecular structures

micelle approximates the properties of a beta fold, promoting formation of a stable complex.

A ribbon-like supramolecular structure may emerge only when the long axes of associating molecules are well aligned with each other. This condition is met when the axial alignment of each unit molecule is determined by its structural elongation (Fig. 1.8).

Another very important property of ribbon-likesupramolecular ligands, promoting complexation, is the exposure of a large hydrophobic surface. The rigid structure and symmetrical distribution of charges in individual molecules prevent

Fig. 1.9 Exposure of non-polar fragments in a ribbon-like micelle composed of self-assembled symmetric molecules (*arrow*)

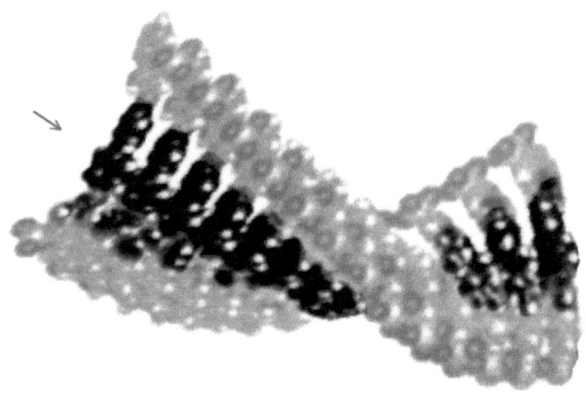

Fig. 1.10 Clouds of CR with solubilized, heat-aggregated IgG molecules (high molecular weight fraction of CR and heat-aggregated IgG complex extracted from molecular sieve chromatography – Sephadex G200). EM imaging (Reproduced by permission of *J. Physiology and Pharmacology*)

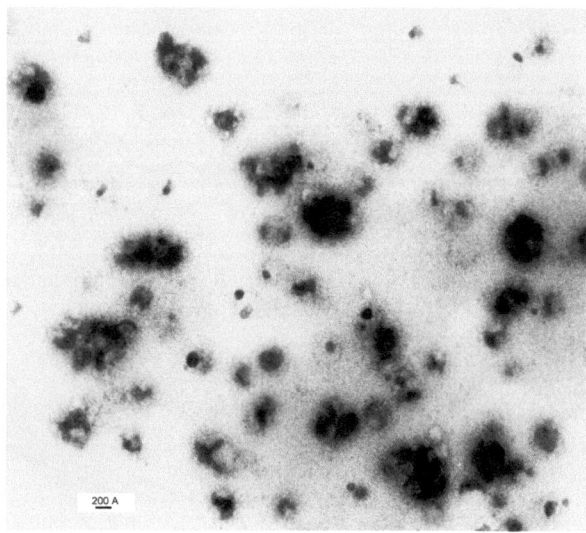

internalization of nonpolar fragments inside the micelle (which occurs in detergents). Such exposure of hydrophobicity on the ligand surface greatly enhances its complexation capabilities, and does so in a specific way: while promoting adhesion, it does not enable the ligand to independently penetrate into the protein – again, in contrast to detergents (Fig. 1.9).

Binding CR increases the protein's polarity, especially in light of the fact that, once anchored, a supramolecular ligand may sometimes propagate beyond the protein and attract additional dye molecules in its environment. As a result, a thermally aggregated protein (such as immunoglobulin G) may persist in solution, surrounded by free dye (Fig. 1.10) [40]. This has been proven through chromatographic separation (on a thin layer of Sephadex G200) of thermally aggregated immunoglobulin G solubilized in complex with the dye. Under EM imaging this heavy fraction appears as a cloud of dye particles with suspended thermally denatured immunoglobulin G, rendered soluble via complexation of CR.

Further adsorption of CR on Sephadex along the column eventually results in precipitation of insufficiently protected immunoglobulin molecules. To enhance the contrast of CR under EM we have added silver ions ($AgNO_3$), which form weak complexes with the dye but remain in solution along the short Sephadex filtration path.

1.2 Structural Adaptability of Molecules Forming Supramolecular Structures

Taken together, the presented characteristics – flat ribbon-like structure, flexibility, large interaction surface and exposure of hydrophobicity – promote interaction and formation of stable complexes with proteins penetrating to areas which are not biologically configured for binding ligands. Another important factor which enhances the adaptability of the CR micelle is some kind of plasticity of individual molecules, permitting rotation about the central bond between aromatic rings, as well as about both lateral azo bonds.

Substituents in conjugated chemical compounds – including polyaromatic compounds – act upon one another. This affects their properties as well as the properties of the entire molecule. The location of such substituents in the molecule is also important. Figure 1.11 presents potentiometric titration of CR and its derivative – 4,4′-bis(1-amino-6-sulphonaphtyl-4-azo-biphenyl Direct Red II), with an identical formula but a different arrangement of polar groups, resulting in a different value of amino groups – pK [41, 42].

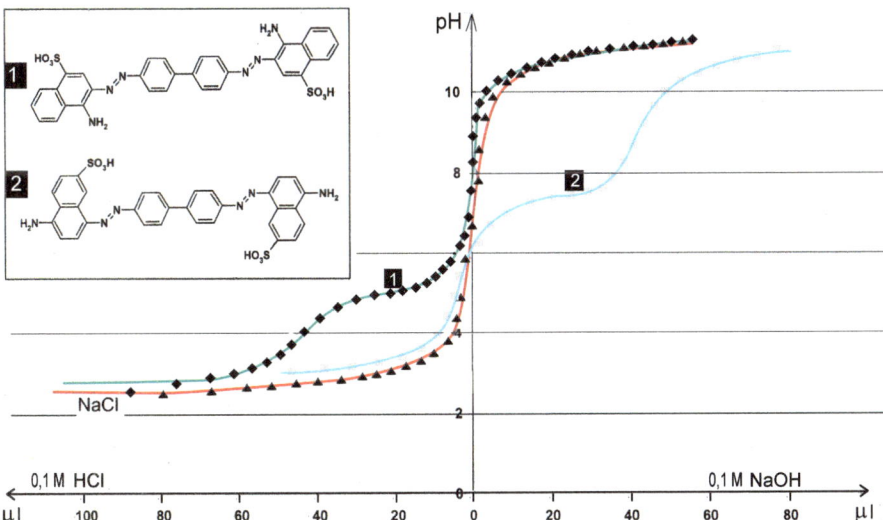

Fig. 1.11 Altered placement of substituents in CR derivative resulting in altered pK of the amino group. Potentiometric titration. Control NaCl - red line

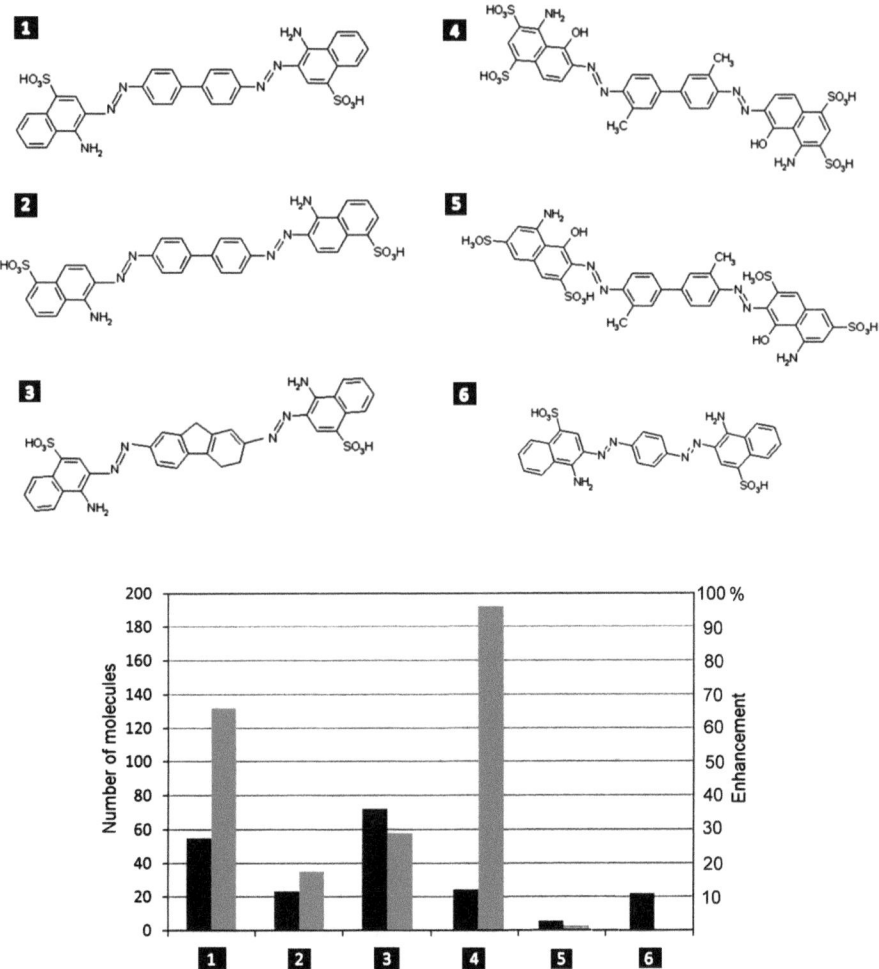

Fig. 1.12 Structural modification of CR molecule – formulas *1, 2, 3, 6* and EB – formulas *4, 5* and their effect on supramolecular binding to proteins: black bars – heat-aggregated IgG (less restrictive binding). Presented value – the number of bound molecules (molar ratio); *gray bars* – native IgG (more restrictive binding – antibodies agglutinating red cells). Presented value – enhancement of agglutination

The rotational freedom associated with azo bonds is strongly dependent on the substituents on aromatic rings; especially those located in close proximity to each azo bond and affecting its polarization, which may either enhance or stifle rotational freedom (Fig. 1.12). To illustrate this fact, we compare CR with its analogue – 4,4′bis(1-amino-5-sulfonaphthyl-2-azo)biphenyl – where greater separation between the azo bonds and the sulfonic groups significantly reduces complexation capabilities (Fig. 1.12 – A and B). In contrast, fixation of the central bond in the fluorene derivative (with the accompanying planarization of the molecule) promotes self-association, but reduces the system's flexibility. This is evidenced by more dense clustering and binding with thermally aggregated immunoglobulin G (where the ligand binding tolerance is high), but reduced capabilities for active site complexation in antibodies and the correspondingly weaker enhancement of agglutination (Fig. 1.12 – *1* and *3*). Further analysis of this phenomenon is possible by measuring enhancement of agglutination in the SRBC-anti-SRBC model caused by complexation of CR. The observed effect is caused by greater involvement of serum polyclonal antibodies in agglutination resulting from complexation of the supramolecular ligand. Complexation capabilities are finally measured by: A – counting the number of dye molecules attached to a single thermally aggregated immunoglobulin G molecule where more than one anchoring site is present; B – assessing the degree to which agglutination is enhanced in the SRBC-anti-SRBC model, under the assumption that the supramolecular ligand increases antibody binding strength and its capacity for immunological complexation. Since such effects require precise alignment of the ligand with the domain V binding site, they provide a measure of the ligand's flexibility. This property can be directly quantified by measuring the readiness for immunological complexation of weak antibodies found in the polyclonal anti-SRBC serum, and the corresponding increase in agglutination.

The relation between degrees of rotational freedom, charge distribution and protein binding capabilities is also evident when comparing EB with Trypan blue (TB). Both dyes differ only with respect to the location of sulfonic groups. In Trypan blue this location is disadvantageous due to its proximity to both the azo bond and the central nonpolar region of the molecule (Fig. 1.12 – *4* and *5*).

The need for a planar ribbon-like micelle becomes clear when we compare the complexation potency of CR with its derivative – 1,4-bis(1-amino-4-sulphonaphtyl-2-azo)phenylene, in which the central biphenyl group has been replaced with a benzene ring, eliminating the need for specific spatial orientation of the molecule's long axis. The resulting micelle is not a ribbon even though its unit molecule closely resembles CR (Fig. 1.12 – *1* and *6*). Instead, it produces a self-associating cylindrical structure, protecting nonpolar fragments from the external environment [43]. This effect also negates the protein complexation capabilities of such a ligand (Fig. 1.13). Comparing the properties of both structurally similar dyes highlights the need for a ribbon-like conformation in supramolecular protein ligands.

Regarding protein structures, supramolecular ligands tend to preferentially form complexes with beta-structure and random coils of polypeptide chains. Elongated non-helical polypeptides represent a good match for the supramolecular ribbon itself, providing the ligand with a convenient anchoring point (Fig. 1.4).

Susceptibility for supramolecular ligand penetration varies from protein to protein. In addition to the degree of packing and the protein's intrinsic stability it also depends – as remarked above – upon function-related conformational rearrangement [35, 36, 38, 39].

The role of most proteins is to interact with specific targets. Such interaction affects the protein itself and often results in partial unfolding, which renders the protein susceptible to further penetration by a supramolecular ligand. This mechanism can be observed e.g. when analyzing the interaction of CR with

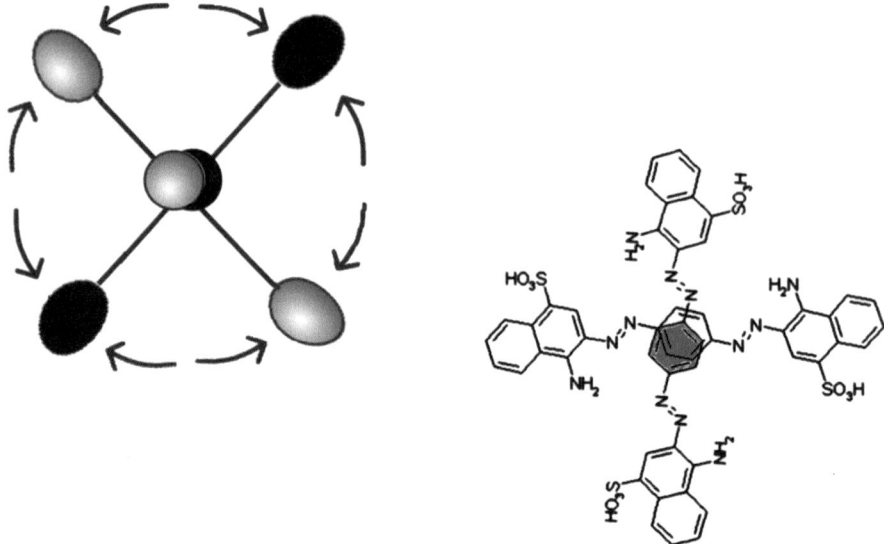

Fig. 1.13 Self-assembly of molecules without imposed orientation (lack of elongated contact area), resulting in a cylindrical supramolecular structure with internalized non-polar fragments. Chemical formula of 1,4-bis(1-1amino-4-sulphonaphtyl-2-azo)phenylene

serum proteins. Note that the bloodstream typically contains acute phase proteins, in complex with their respective ligands, and that such complexes are recognized and eliminated by macrophages and liver enzymes. We may therefore suspect that the capability for such selective elimination depends on function-related structural changes which occur in proteins introducing some local instability.

1.3 Specificity of Congo Red Complexation

The complexation capabilities of CR increase along with the dye's concentration. This is related to increased probability of penetration into proteins as they undergo dynamic – and often temporary – structural changes. Furthermore, increased concentrations favor supramolecular association, resulting in longer micelles with more pronounced dipole characteristics. This effect can be revealed by measuring electrophoretic migration distance on the electrophoretic plate and present it as a function of dye concentration.

It seems that the properties of CR change qualitatively as its concentration increases, favoring protein complexation. At high concentrations the dye is even capable of penetrating into proteins at room temperature. The application of DMSO results in dissociation of the supramolecular structure; consequently, the electrophoretic migration speed again becomes independent of concentration (Fig. 1.14).

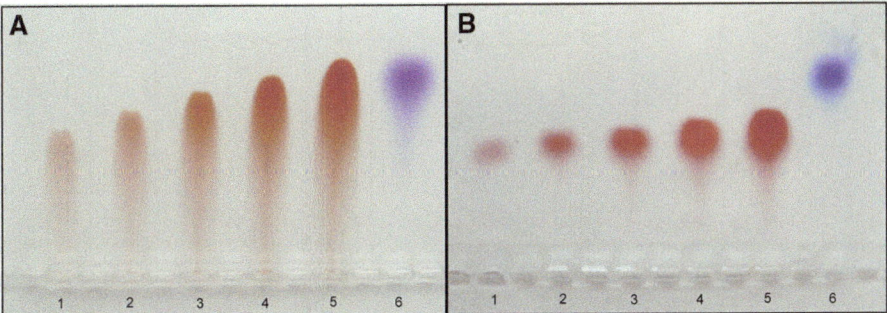

Fig. 1.14 Increased electrophoretic migration rate of CR corresponding to increased dye concentration: **A** - 1,2,3,4,5, position 6 - bromophenol blue dye. **B** - concentration-independent migration in (DMSO – buffer mixture 1:2). Evidence of close CR self-assembly

Fig. 1.15 Two mutually-related complexes consisting of L-lambda chains and CR molecules, exhibiting different migration rates in electrophoresis (*b* and *c*) due to different well defined dye load. *a* – L chain, *b* – L chain complexed with 4 molecular CR ligand, *c* – L chain complexed with 5–8 molecular CR ligand, *d* – CR excess. Complexation induced by the step-wise increased concentration of CR - 1,2,3,4

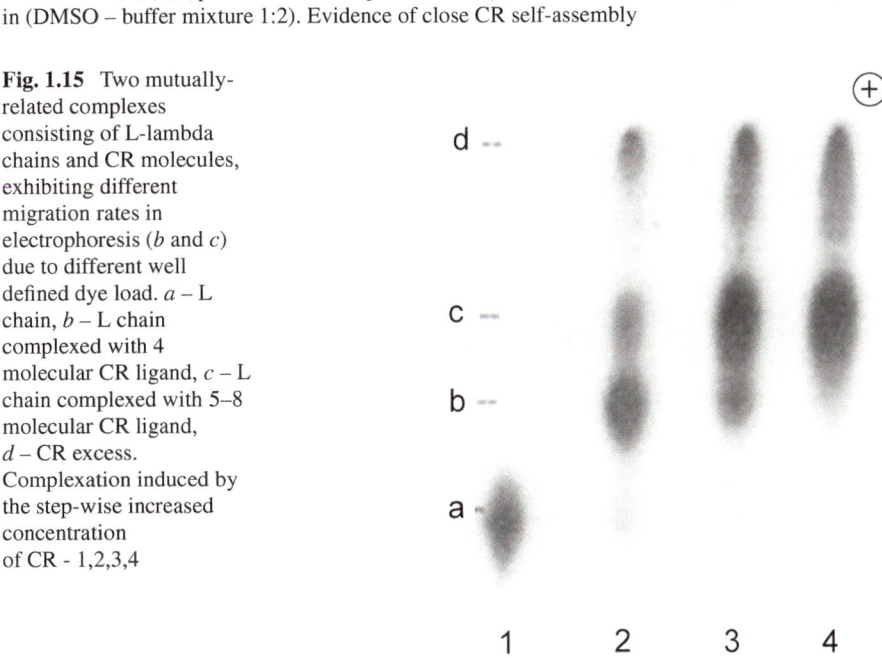

This proves that concentration determines the emergent, supramolecular properties of the dye – particularly its capability to form stable complexes with proteins.

In order to demonstrate this phenomenon, we have selected an immunoglobulin light chain which is relatively resistant to CR complexation in its native form. Heating the protein promotes complexation, but even at room temperature high concentrations result in two distinct complexes which migrate faster than the base protein – depending on the number of ligand molecules present in each complex. Here, complexation involves the variable V domain (Fig. 1.15) [44]. The "slow" migrating fraction carries ligands composed of four dye molecules, while in the "fast" fraction

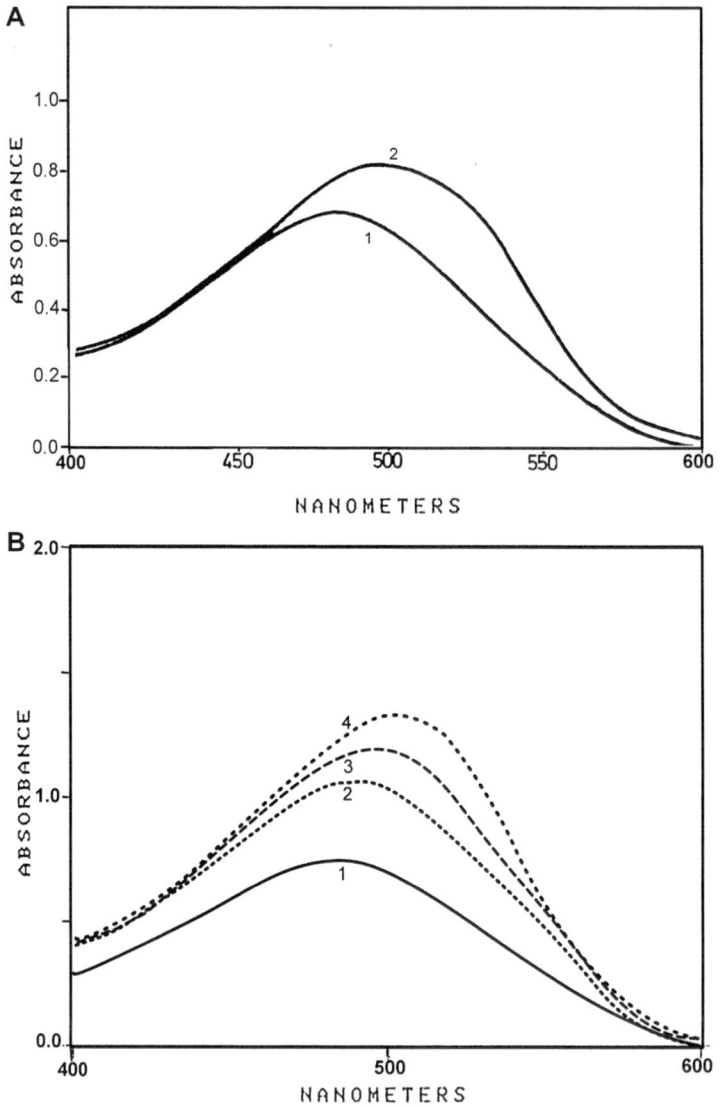

Fig. 1.16 Spectral shift of CR in a low-polarity environment – **A** -*1* – native concanavalin, *2* – heated concanavalin) and **B** - in alcohols (*1*- native concanavalin *2*-methanol, *3*-ethanol, *4*-propanol) characterized by varying polarity

the size of the ligand varies between 5 and 8 molecules. Notably, the "fast" fraction produces a smeared electrophoretic band, showing that the ligand grows over time and that larger complexes are produced with greater difficulty than smaller ones.

The supramolecular ligand forms a tight bond with the protein and adapts itself to the new environment, as evidenced by a spectral shift towards greater wavelengths. This effect results from transitioning between water and an environment characterized by lower values of the dielectric constant (Fig. 1.16A) [45, 46].

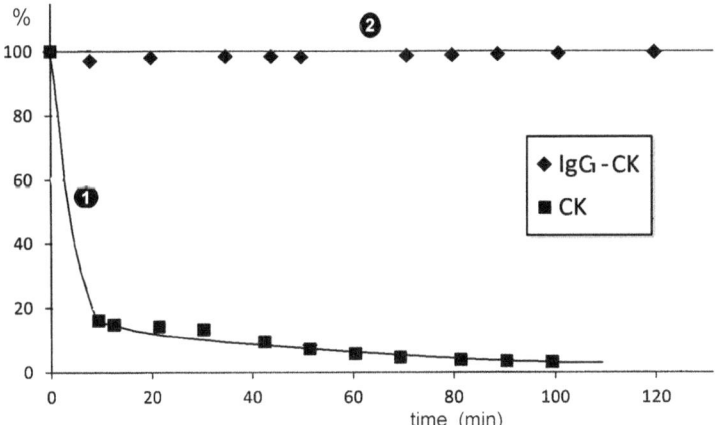

Fig. 1.17 Dye protected against acidification by complexed protein measured by spectral change. *1* – free dye, *2* – dye bound to protein (L chain IgG)

Fig. 1.18 Dye protected against reduction by complexed protein measured by spectral change. *1* – dye bound to protein, *2* – free dye

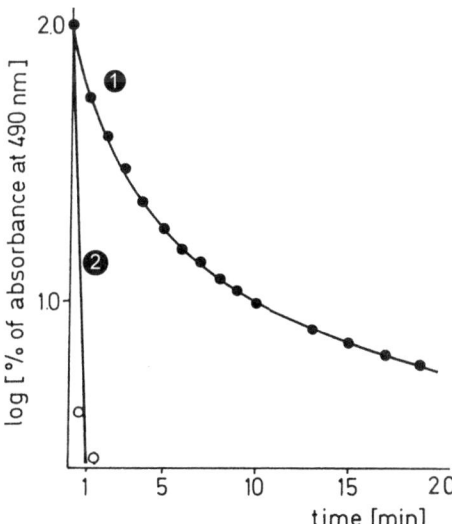

To further illustrate this effect, the spectrum of CR has been analyzed in the presence of alcohols containing increasingly larger nonpolar components: methanol, ethanol and propanol. The observed shift towards greater wavelengths confirms the stated hypothesis (Fig. 1.16B).

A complexed ligand bound in protein interior is protected against acidification by the protein. CR changes its color in an acidic environment due to change in the ionization of its amino groups. The color shift from red to blue is decisive and rapid in unbound dye, whereas dye-protein complexes retain their red coloration for some time (Fig. 1.17). Reduction by sodium dithionite deprives the dye of its color due to cleavage of azo bonds; however, protein complexation slows this reaction down significantly (Fig. 1.18).

Fig. 1.19 Self-assembly tendency correlated with corresponding different dye complexation activity measured as the yield of dye protein complexation at increasing temperature. Lines *1* and *2* according to chemical formulas *1* and *2* respectively. 1 - CR, 2 - 1,4-bis(1-1amino-4-sulphonaphtyl-2-azo) phenylene

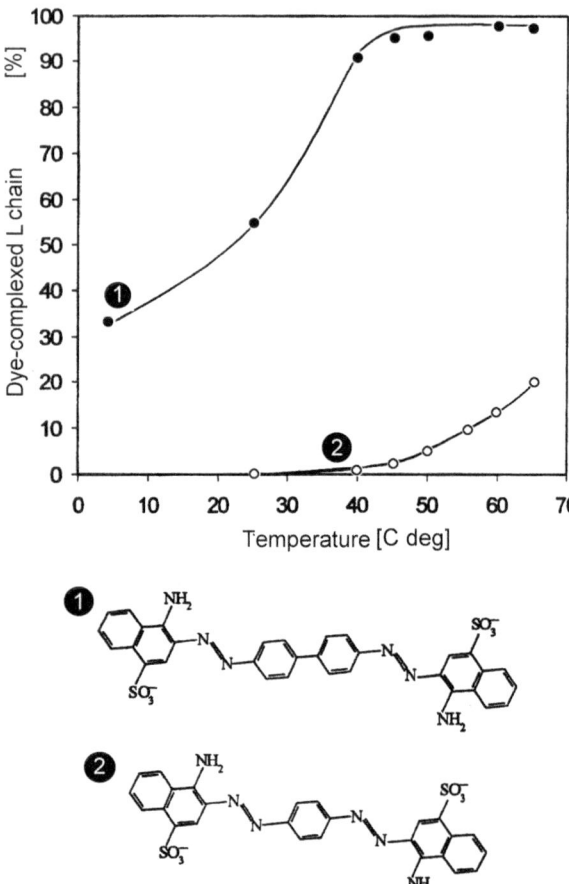

In summary, complexes consisting of supramolecular dyes and proteins appear to result from penetration of associated dye molecules into the target protein. The capability for such penetration depends on the cohesiveness of the dye micelle, as well as its shape.

In order for a supramolecular aggregation to function as a distinct protein ligand, the unit composed of self-assembled molecules must behave as a coherent whole. This depends on the self-association potency of the target substance – powerful self-association produces a ligand which readily interacts with proteins (Figs. 1.19 and 1.20).

Another important property of supramolecular dyes is their capability to intercalate foreign bodies (other than the self-associating unit molecules), resulting in ligands which can introduce foreign substances into proteins even when the protein does not, by itself, react with such substances [17, 47]. Rhodamine B – a basic dye which exhibits strong fluorescence and is therefore useful in imaging studies – may be intercalated into CR micelles and bound to proteins. Other potential intercalants

Fig. 1.20 Efficiency of formation the protein-dye complex corresponding to self-assembly tendency – registered at increasing temperatures. Lines *1* and *2* correspond to chemical molecules of chemical formulas 1 (EB) and 2 respectively (TB)

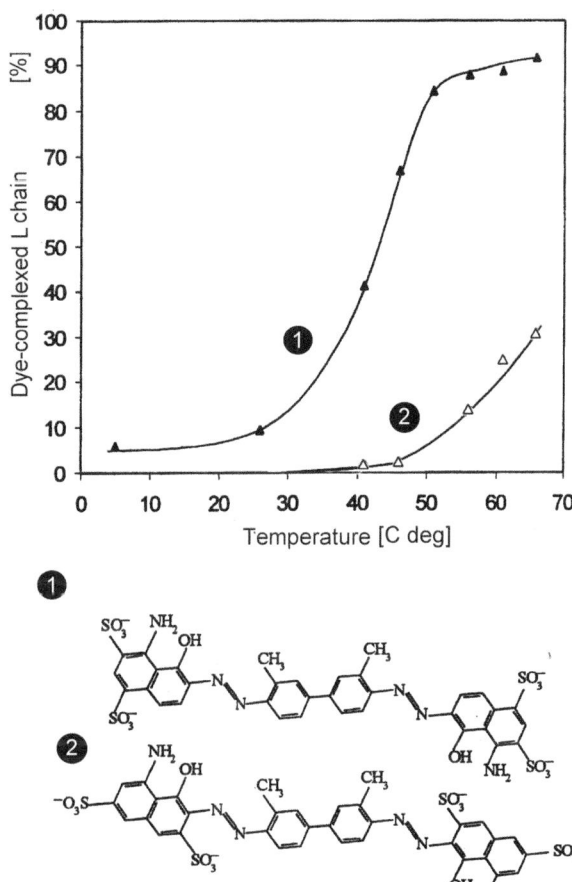

include heavy metal ions – such as in the case of TY, used as a carrier for silver ions to provide contrast for EM imaging of amyloid deposits [48]. The same mechanism may be used to introduce some alterations to properties of proteins.

Acknowledgements Work financially supported by Collegium Medicum – Jagiellonian University grant system – grant # K/ZDS/006363.

References

1. Kay LE (1998) Protein dynamics from NMR. Biochem Cell Biol 76(2–3):145–152
2. Doyle DA, Lee A, Lewis J, Kim E, Sheng M, MacKinnon R (1996) Crystal structures of a complexed and peptide-free membrane protein-binding domain: molecular basis of peptide recognition by PDZ. Cell 85(7):1067–1076

3. Fuentes EJ, Der CJ, Lee AL (2004) Ligand-dependent dynamics and intramolecular signalling in a PDZ domain. J Mol Biol 335(4):1105–1115
4. Fraser JS, Clarkson MW, Degnan SC, Erion R, Kern D, Alber T (2009) Hidden alternative structures of proline isomerase essential for catalysis. Nature 462(7273):669–673
5. McLaughlin RN Jr, Poelwijk FJ, Raman A, Gosal WS, Ranganathan R (2012) The spatial architecture of protein function and adaptation. Nature 491(7422):138–142
6. Laskowski RA, Gerick F, Thornton JM (2009) The structural basis of allosteric regulation in proteins. FEBS Lett 583(11):1692–1698
7. Gunasekaran K, Ma B, Nussinov R (2004) Is allostery an intrinsic property of all dynamic proteins? Proteins 57(3):433–443
8. Kern D, Zuiderweg ER (2003) The role of dynamics in allosteric regulation. Curr Opin Struct Biol 13(6):748–757
9. Tompa P (2011) Unstructural biology coming of age. Curr Opin Struct Biol 21(3):419–425
10. Wright PE, Dyson HJ (1999) Intrinsically unstructured proteins: re-assessing the protein structure-function paradigm. J Mol Biol 293(2):321–331
11. England JL (2011) Allostery in protein domains reflects a balance of steric and hydrophobic effects. Structure 19(7):967–975
12. Koshland DE Jr (1959) Enzyme flexibility and enzyme action. J Cell Comp Physiol 54:245–258
13. Zeng C, Chen Y, Kirschbaum K, Lambright KJ, Jin R (2016) Emergence of hierarchical structural complexities in nanoparticles and their assembly. Science 354(6319):1580–1584
14. Liu W, Tagawa M, Xin HL, Wang T, Emamy H, Li H, Yager KG, Starr FW, Tkachenko AV, Gang O (2016) Diamond family of nanoparticle superlattices. Science 351(6273):582–586
15. Sacanna S, Irvine WT, Chaikin PM, Pine DJ (2010) Lock and key colloids. Nature 464(7288):575–578
16. Desiraju GR (2001) Chemistry beyond the molecule. Nature 412(6845):397–400
17. Swanson BD, Sorensen LB (1995) What forces bind liquid crystals? Phys Rev Lett 75(18):3293–3296
18. Herzfeld J (1996) Entropically driven order in crowded solutions: from liquid crystals to cell biology. Acc Chem Res 29(1):31–37
19. Lv JA, Liu Y, Wei J, Chen E, Qin L, Yu Y (2016) Photocontrol of fluid slugs in liquid crystal polymer microactuators. Nature 537(7619):179–184
20. Evers CH, Luiken JA, Bolhuis PG, Kegel WK (2016) Self-assembly of microcapsules via colloidal bond hybridization and anisotropy. Nature 534(7607):364–368
21. Skowronek M, Stopa B, Konieczny L, Rybarska J, Piekarska B, Szneler E, Bakalarski G, Roterman I (1998) Self-assembly of Congo red – a theoretical approach to identify its supramolecular organization In water and salt solutions. Biopolymers 46:267–281
22. Król M, Roterman I, Piekarska B, Konieczny L, Rybarska J, Stopa B, Spólnik P, Szneler E (2005) An approach to understand the complexation of supramolecular dye Congo red with immunoglobulin L chain lambda. Biopolymers 77(3):155–162
23. Stopa B, Rybarska J, Drozd A, Konieczny L, Król M, Lisowski M, Piekarska B, Roterman I, Spólnik P, Zemanek G (2006) Albumin binds self-assembling dyes as specific polymolecular ligands. Int J Biol Macromol 40(1):1–8
24. Edelman GM, Gally JA (1962) The nature of Bence-Jones proteins. Chemical similarities to polypetide chains of myeloma globulins and normal gamma-globulins. J Exp Med 116:207–227
25. Nakano T, Matsui M, Inoue I, Awata T, Katayama S, Murakoshi T (2011) Free immunoglobulin light chain: its biology and implications in diseases. Clin Chim Acta 412(11–12):843–849
26. Leitzgen K, Knittler MR, Haas IG (1997) Assembly of immunoglobulin light chains as a prerequisite for secretion. A model for oligomerization-dependent subunit folding. J Biol Chem 272(5):3117–3123
27. Kaplan B, Livneh A, Sela BA (2011) Immunoglobulin free light chain dimers in human diseases. Sci World J 11:726–735

28. Charafeddine KM, Jabbour MN, Kadi RH, Daher RT (2012) Extended use of serum free light chain as a biomarker in lymphoproliferative disorders: a comprehensive review. Am J Clin Pathol 137(6):890–897
29. Woodcock S, Henrissat B, Sugiyama J (1995) Docking of congo red to the surface of crystalline cellulose using molecular mechanics. Biopolymers 36(2):201–210
30. Khurana R, Gillespie JR, Talapatra A, Minert LJ, Ionescu-Zanetti C, Millett I, Fink AL (2001) Partially folded intermediates as critical precursors of light chain amyloid fibrils and amorphous aggregates. Biochemistry 40(12):3525–3535
31. Howie AJ, Brewer DB (2009) Optical properties of amyloid stained by Congo red: history and mechanisms. Micron 40(3):285–301
32. Buell AK, Dobson CM, Knowles TP, Welland ME (2010) Interactions between amyloidophilic dyes and their relevance to studies of amyloid inhibitors. Biophys J 99(10):3492–3497
33. Wang Y, Liu Y, Deng X, Cong Y, Jiang X (2016) Peptidic β-sheet binding with Congo Red allows both reduction of error variance and signal amplification for immunoassays. Biosens Bioelectron 86:211–218
34. Frid P, Anisimov SV, Popovic N (2007) Congo red and protein aggregation in neurodegenerative diseases. Brain Res Rev 53(1):135–160
35. Lendel C, Bolognesi B, Wahlström A, Dobson CM, Gräslund A (2010) Detergent-like interaction of Congo red with the amyloid beta peptide. Biochemistry 49(7):1358–1360
36. Rybarska J, Konieczny L, Roterman I, Piekarska B (1991) The effect of azo dyes on the formation of immune complexes. Arch Immunol Ther Exp 39(3):317–327
37. Jagusiak A, Konieczny L, Krol M, Marszalek P, Piekarska B, Piwowar P, Roterman I, Rybarska J, Stopa B, Zcmanck G (2015) Intramolccular immunological signal hypothcsis rcvivcd-structural background of signalling revealed by using Congo Red as a specific tool. Mini Rev Med Chem 4(13):1104–1113
38. Weber K, Osborn M (1969) The reliability of molecular weight determinations by dodecyl sulfate-polyacrylamide gel electrophoresis. J Biol Chem 244(16):4406–4412
39. Kaszuba J, Konieczny L, Piekarska B, Roterman I, Rybarska J (1993) Bis-azo dyes interference with effector activation of antibodies. J Physiol Pharmacol 44(3):233–242
40. Shrestha S, Shim YS, Kim KC, Lee KH, Cho H (2004) Evans Blue and other dyes as protein tyrosine phosphatase inhibitors. Bioorg Med Chem Lett 14(8):1923–1926
41. Piekarska B, Konieczny L, Rybarska J, Stopa B, Spólnik P, Roterman I, Król M (2004) Intramolecular signaling in immunoglobulins – new evidence emerging from the use of supramolecular protein ligands. J Physiol Pharmacol 55(3):487–501
42. Stopa B, Piekarska B, Konieczny L, Rybarska J, Spólnik P, Zemanek G, Roterman I, Król M (2003) The structure and protein binding of amyloid-specific dye reagents. Acta Biochim Pol 50(4):1213–1227
43. Spólnik P, Konieczny L, Piekarska B, Rybarska J, Stopa B, Zemanek G, Król M, Roterman I (2004) Instability of monoclonal myeloma protein may be identified as susceptibility to penetration and binding by newly synthesized Congo red derivatives. Biochimie 86(6):397–401
44. Zemanek G, Konieczny L, Piekarska B, Rybarska J, Stopa B, Spólnik P, Urbanowicz B, Nowak M, Król M, Roterman I (2002) Egg yolk platelet proteins from Xenopus laevis are amyloidogenic. Folia Histochem Cytobiol 40(3):311–318
45. Piekarska B, Konieczny L, Rybarska J, Stopa B, Zemanek G, Szneler E, Król M, Nowak M, Roterman I (2001) Heat-induced formation of a specific binding site for self-assembled Congo Red in the V domain of immunoglobulin L chain lambda. Biopolymers 59(6):446–456
46. Piekarska B, Konieczny L, Rybarska J, Stopa B, Zemanek G, Szneler E, Król M, Nowak M, Roterman I (2001) Heat-induced formation of a specific binding site for self-assembled Congo Red in the V domain of immunoglobulin L chain lambda. Biopolymers 9(6):446–456

47. Konieczny L, Piekarska B, Rybarska J, Stopa B, Krzykwa B, Noworolski J, Pawlicki R, Roterman I (1994) Bis azo dye liquid crystalline micelles as possible drug carriers in immuno-targeting technique. J Physiol Pharmacol 45(3):441–454
48. Konieczny L, Piekarska B, Rybarska J, Skowronek M, Stopa B, Tabor B, Dabroś W, Pawlicki R, Roterman I (1997) The use of Congo red as a lyotropic liquid crystal to carry stains in a model immunotargeting system – microscopic studies. Folia Histochem Cytobiol 35(4):203–210

Chapter 2
Supramolecular Congo Red as Specific Ligand of Antibodies Engaged in Immune Complex

Anna Jagusiak, Joanna Rybarska, Barbara Piekarska, Barbara Stopa, and Leszek Konieczny

Abstract Supramolecular Congo red has been used to validate long-lasting theories regarding intramolecular signaling in antibodies and its relation to activation of the complement system. Strong enhancement of antigen-antibody complexation resulting from the binding of supramolecular ligands enables also polyclonal antibodies having intermediate affinity to trigger complement cascade apart of high affinity antibody fraction. This would not have been possible in the absence of Congo red. The property of antibodies provides specifically their ability to trigger the complement system allowed when sufficient structural strain is produced by antigen complexation provides an evidence of intramolecular signaling.

The selective complexation of supramolecular ligands with antibodies engaged in immune complexes enables their using as carriers of drugs in immunotargeting system.

Keywords Intramolecular immunological signal • Complement activation • IgG V domain stability • N-terminal fragment • Enhancement of antigen binding • Congo red as carrier of drugs • Immunotargeting system • Congo red selective complexation of antibodies in immune complexes

Self-associating organic molecules which form ribbonlike micellar structures may, owing to their structural characteristics, penetrate inside proteins and form stable complexes. Such penetration is possible in areas of the protein which have been destabilized, either temporarily or permanently – such as antibody/antigen complexes. Since Congo red (CR) has been used in research as the most typical

A. Jagusiak (✉) • J. Rybarska • B. Piekarska • B. Stopa • L. Konieczny
Chair of Medical Biochemistry, Jagiellonian University – Medical College,
Kopernika 7, 31-034 Krakow, Poland
e-mail: anna.jagusiak@uj.edu.pl; mbstylin@cyf-kr.edu.pl; mbpiekar@cyf-kr.edu.pl;
barbara.stopa@uj.edu.pl; mbkoniec@cyf-kr.edu.pl

© The Author(s) 2018
I. Roterman, L. Konieczny (eds.), *Self-Assembled Molecules – New Kind of Protein Ligands*, https://doi.org/10.1007/978-3-319-65639-7_2

supramolecular protein ligand, the presented experiments and analysis also focus on this particular dye. CR binds strongly to antibodies, enabling us to study (among others) intramolecular signaling related to complement system activation. What is more, the mutual affinity of CR and immune complexes paves the way towards immunotargeting, i.e. targeted delivery of drugs. This is due to the fact that supramolecular CR – a micelle-like structure – may intercalate foreign bodies, including drug molecules. Congo red does not react with free antibodies – it is only capable of binding to antibody/antigen complexes where structure of antibody undergoes some alteration due to interaction with the antigen. Any potential drugintercalated into the CR micelle can thus be delivered to an area where the antigen is plentiful, ensuring targeted action. This chapter discusses the presented topics in detail.

2.1   Looking for Evidence of Postulated Intra-molecular Immunological Signaling

Once the structure of immunoglobulins has been divined, it soon became clear that their Fab and Fc fragments play differing roles in the process of triggering immuno-logical response. While the Fab fragment selectively binds to the antigen, the Fc fragment – separated by a hinge – appears to be involved in triggering complement system activation through complexation of the C1q subcomponent. Notably, the Fab-antigen interaction is independent of Fc and proceeds even when the Fc frag-ment has been removed by digestion [1, 2].

The complement system is a collection of proteins which attack and destroy cells recognized as alien by the immune complex. Strict control over this mechanism is critical for homeostasis and therefore represents an important study subject in medical research. In accordance with prevalent views, such control is maintained by intramolecular rearrangements which carry information from Fab to Fc, and then onwards to C1q (Fig. 2.1).

Fig. 2.1 Schematic depiction of the intramolecular signaling pathway inside the antibody (*dashed line*)

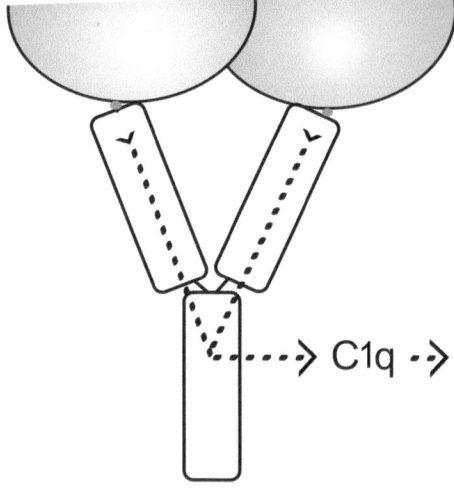

Nevertheless, despite significant effort by many leading researchers, the specifics of this mechanism have proven exceedingly difficult to elucidate and some uncertainties persist. In attempting to explain intramolecular signaling, analysts initially focused on the hinge region which links Fab and Fc. Experimental data indicates that subclasses of immunoglobulins which differ with respect to the composition of this hinge region also exhibit variable efficiency of Fab-to-Fc signal transmission, and moreover that reduction of the disulfide bond in the hinge region prevents successful activation of the complement system [3–5]. In turn, some attention was directed towards structural strain in the antibody molecule, produced by antigen binding and regarded as a possible signal carrier. This view is embodied in the so-called distortive mechanism theory. Another competing theory proposed an "all or none" switching mechanism, i.e. an allosteric model based on the assumption that immunoglobulins are, in fact, allosteric [6].

Since none of the presented models succeeded in providing a satisfactory explanation, further analysis was needed. Some researchers noted the fact that, under ordinary circumstances, the formation of an active immune complex involves many different antibodies, and that complement activation requires local concentration of Fc fragments. This so-called associative model appeared to explain the signaling puzzle to a sufficient degree, particularly given the lack of evidence favoring intramolecular signaling [7, 8]. Earlier theories were swept aside and the issue appeared solved. This situation persisted for many years, until scientists learned how to produce monoclonal antibodies via crystallization of Fab fragments cleaved from the IgG molecule, and formulated new analysis protocols based on the use of small antigens (haptens) [9–11]. Surprisingly, these studies produced little in the way of useful results. Structural changes appeared small, even negligible – again suggesting that intramolecular signaling must somehow involve torsional effects, which emerge only when the antigen is bound to a complete, two-arm antibody.

Spectacular progress in genetics achieved in the 1980s, particularly the ability to synthesize arbitrarily modified antibodies, brought new hope of understanding the purported intramolecular signaling mechanism. Still however, despite some focus on the interaction of CH_1 and CH_2 domains, the problem of signal remained practically unsolved [12–14].

Our research group decided to attack the problem through chemical recombination of antibodies by digestion, reduction and re-joining of immunoglobulin fragments solely by disulfide bonds. The goal was to determine whether this kind of modified structure would retain the ability to carry the signal to Fc, despite major alterations in the hinge region. This process is illustrated in Fig. 2.2.

While the "full" two-arm molecule constructed from free Fab and Fc fragments exhibited complement activation potential (to a limited degree), its one-arm equivalent proved entirely inert. This suggested that even a deficient antibody may transmit the signal, if only suitable conditions exist for structural strain to emerge. Nevertheless, the participation of the hinge region in signal transmission remained a mystery [3, 15, 16].

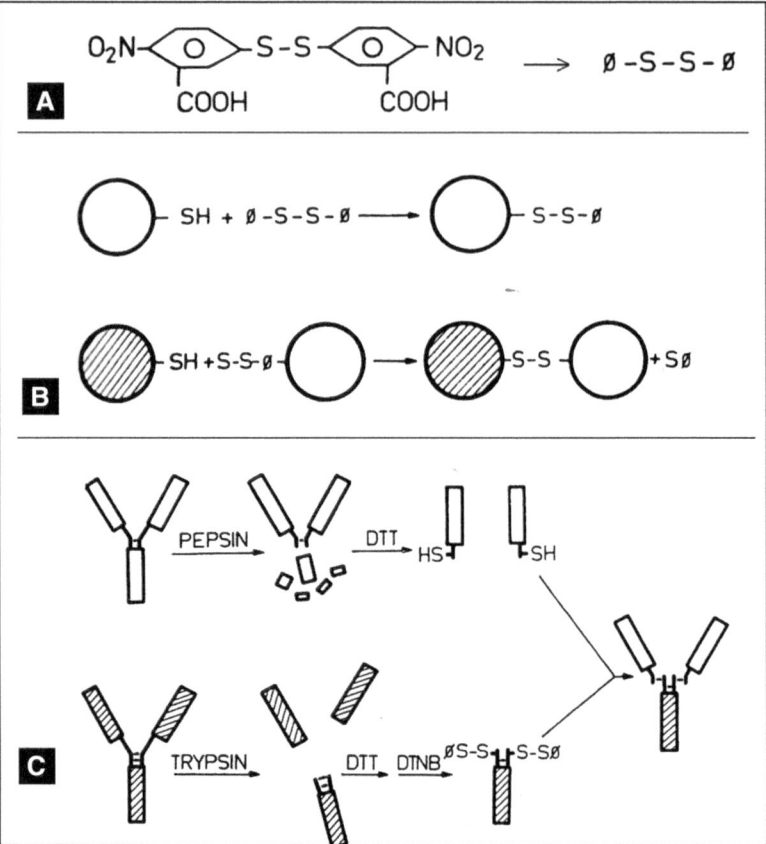

Fig. 2.2 Schematic view: (**A** and **B**) Controlled formation of linking disulfide bonds. (**C**) Production of recombinant IgG

2.2 Evidences of Intramolecular Signaling Supplied by Using Congo Red

A whole new approach to the problem was enabled by the use of CR, based on our team's original concept. While CR had long been known as a useful amyloid stain, its interaction with amyloids was explained as individual molecules attaching themselves to specific binding sites which recognize the dye. In contrast, our study revealed that CR may form complexes with a wide variety of proteins and that it does so as a supramolecular ligand – i.e. a distinct structure consisting of many associated dye molecules acting as a single unit [17–25].

Fig. 2.3 V domain of the
L chain lambda, with its
hightlighted N-terminal
fragment covering gap
created by its removal.
Space filling model

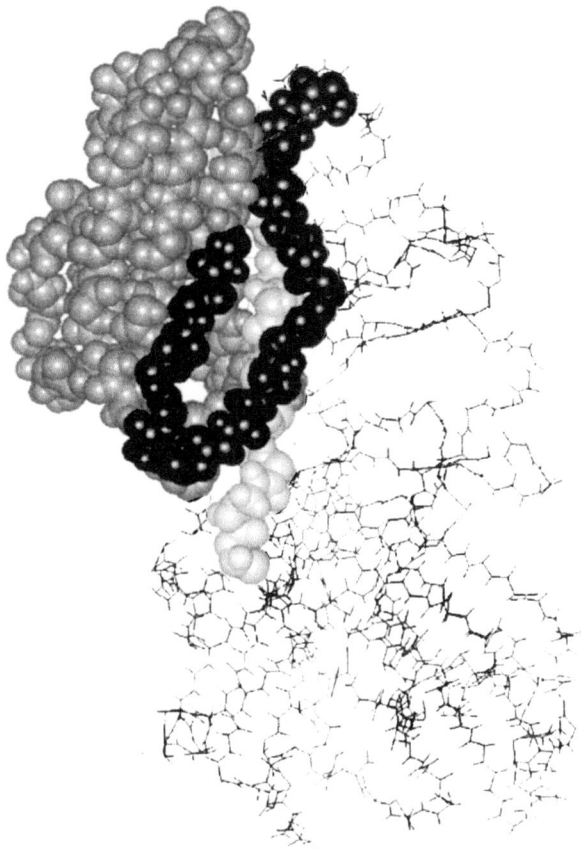

An important breakthrough occurred when CR was found to interact with immune complexes, but not with free antibodies. This phenomenon shed new light on the intramolecular signaling pathways leading to complement system activation. Such selective binding suggested that CR is capable of anchoring itself in the V domain of the antibody, which also happens to be the site of the greatest structural strain resulting from antigen complexation. Furthermore, the V domain also houses the N-terminal fragment, which, by default, is relatively unstable (Fig. 2.3) [26–28].

Confirmation of this theory was provided by analyzing the complexation potential of CR vs. light chain dimers progressively destabilized through heating or increased concentrations of the dye. The displacement of the N-terminal fragment from its packing locus "opens up" the V domain, enabling the supramolecular dye to penetrate and anchor itself in its interior. Under experimental conditions, the first complexes to emerge involve ligands composed of four molecules. As the dye concentration increases, the ligand may grow to include up to eight dye molecules (Fig. 2.4A, B respectively). This is evidenced by electrophoresis, where larger com-

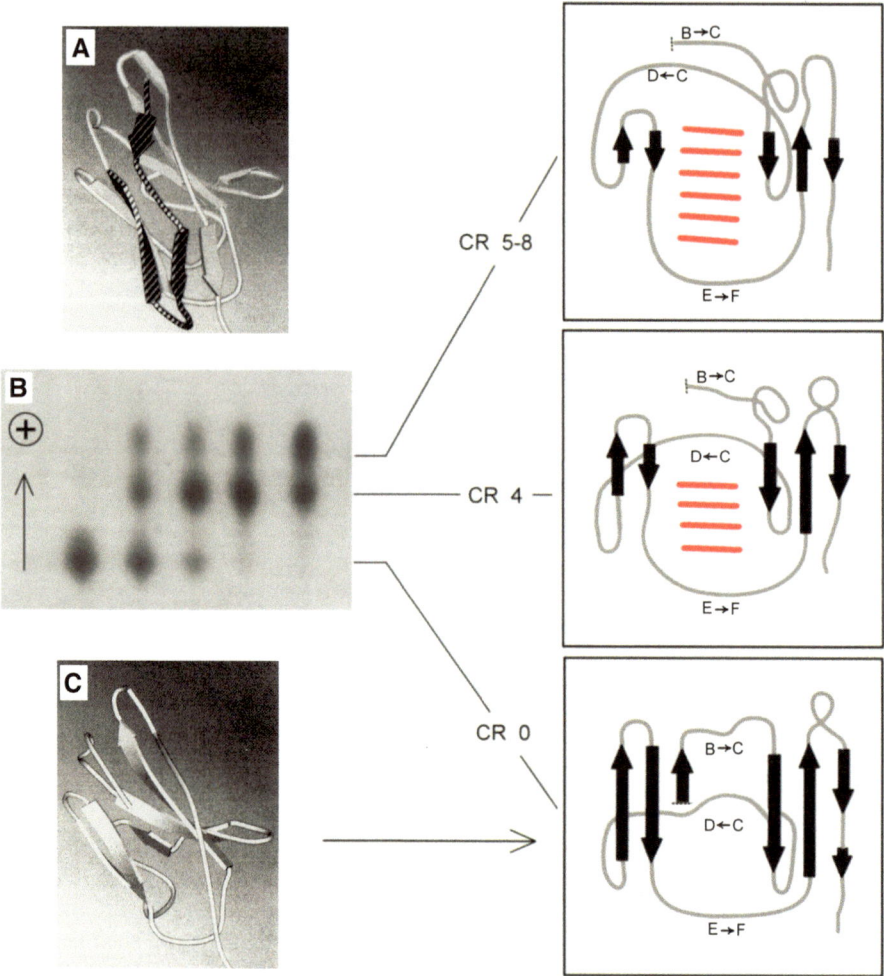

Fig. 2.4 Complexation of L chain dimer with CR (**A**) Light chain V domain. (**B**) Thermally generated complexes of CR with the L chain V domain presented by corresponding models. (**C**) V domain deprived of its N-terminal fragment through digestion. Agarose electrophoresis of L chains induced to form complexes with CR upon the stepwise increasing temperature

plexes migrate faster due to their greater charge (contributed by CR). In each case, the trigger for complexation appears to be the N-terminal fragment, which is displaced from its packing locus and replaced by the supramolecular ligand. The displaced N-terminal fragment subsequently becomes susceptible to digestion [26, 29–31].

CR/light chain complexation may be accelerated by heating, which further perturbs the N-terminal fragment. In contrast, when dealing with immune complexes, complexation appears to be induced by structural strain resulting from antigen bind-

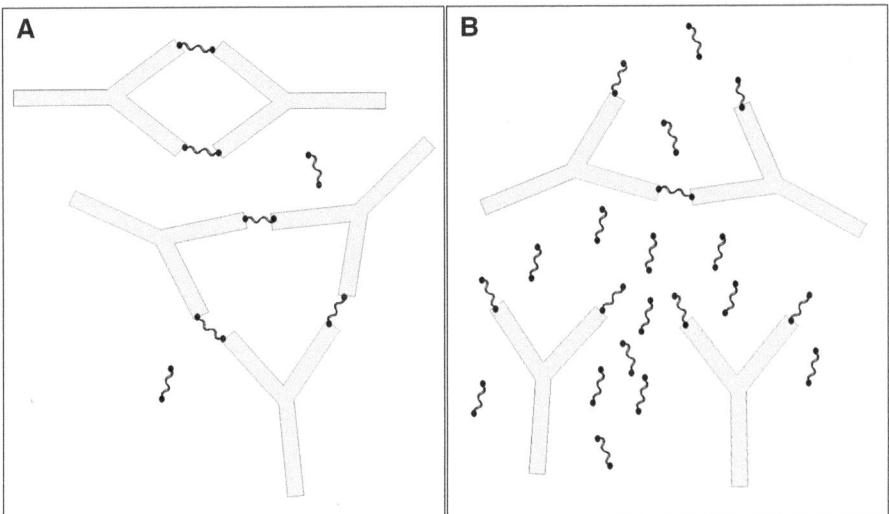

Fig. 2.5 Immune complexation of antibodies linked by binary haptens (**A**) and breakage of immune complexes caused by overabundance of hapten particles (**B**)

ing. This phenomenon, however, only emerges in complete two-arm antibodies, attaching themselves to antigen determinants located randomly on the cell surface. Neither isolated Fab fragments nor their dimers are capable of binding CR, even in their complexed state. This proves that the antibody-antigen reaction is not directly responsible for the affinity to CR, and that dye complexation requires structural strain in the antibody molecule [26, 29–31].

CR is complexed by whole antibodies in complex with haptens, but only when they are fixed on a solid surface, i.e. under conditions which lead to structural strain in bivalent antibodies.

CR also binds to anti-TNP antibodies linked by a hapten if the hapten itself is bivalent and creates strain in the antibodies it links (e.g. oxidized glutathione with amino groups substituted with TNP – Fig. 2.5).

This can be confirmed under electrophoresis, since the antibodies bound by the binary hapten form immune complexes, become soluble and migrate more rapidly when treated with CR. Their number grows as the concentration of the hapten increases, eventually reaching a maximum beyond which a falloff is expected (Fig. 2.6) – due to the fact that when the hapten is overly abundant, it becomes monovalent and therefore produces no strain in the attached antibodies (Fig. 2.5) [32].

The link between CR and immune complexes exhibits one more remarkable property: **it turns out that complexation of CRgreatly enhances theantigen/ antibodycomplexation capabilities**. This is evidenced by a significant increase in the fraction of antibodies involved in immune complexes, especially in relation to low-affinity antibodies which are naturally present in the polyclonal serum. It should be noted that the serum contains many different types of antibodies with

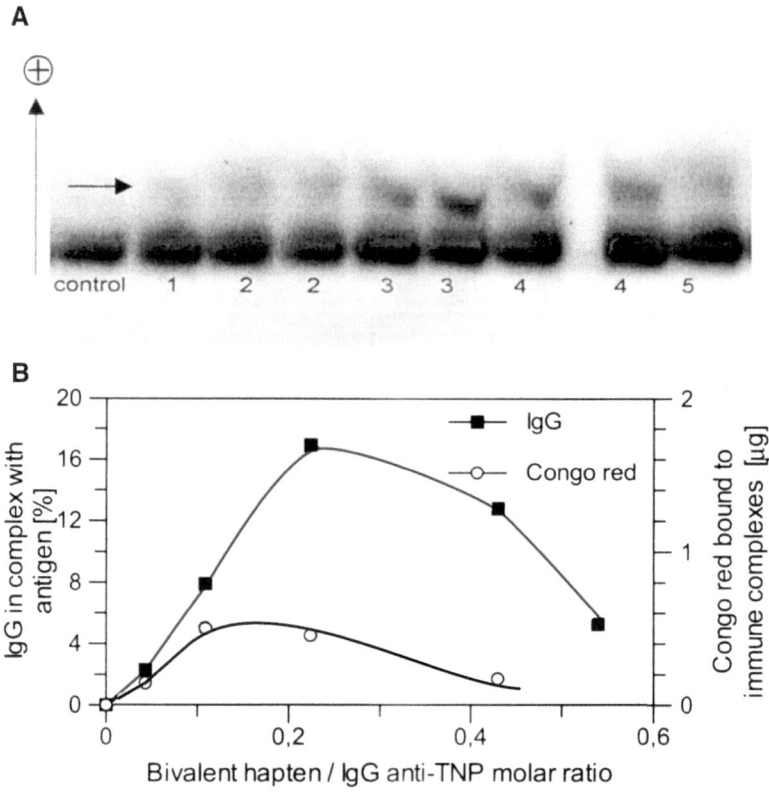

Fig. 2.6 Experimental evidence of formation soluble complexes with CR by binary hapten-linked antibodies (as depicted in Fig. 2.5). *Left to right*: increasing hapten concentrations. The mobile fraction (indicated by the *arrow*) comprises soluble immune complexes which have gained the ability to bind CR (Reproduction by permission – *J Physiology and Pharmacology*)

varying degrees of affinity – due to the inherent randomness in the antibody synthesis process. Low-affinity antibodies cannot form stable immune complexes and are washed out in the absence of CR. When the dye is present, their ability to bind antigens increases and the number of immune complexes per unit of volume grows (Fig. 2.7 **A, B**).

At this point it would be useful to determine the mechanism which drives increased complexation capabilities in the presence of CR, and also to find out whether such upregulation is accompanied by the corresponding increase in complement system activity (which would prove the existence of an intramolecular signal).

The assumption that, by binding its natural biological ligand, the protein undergoes structural rearrangement which favors penetration and complexation of supramolecular dye would explain why the ligand cannot be easily released once the protein-dye complex has formed. This phenomenon appears actual in situations

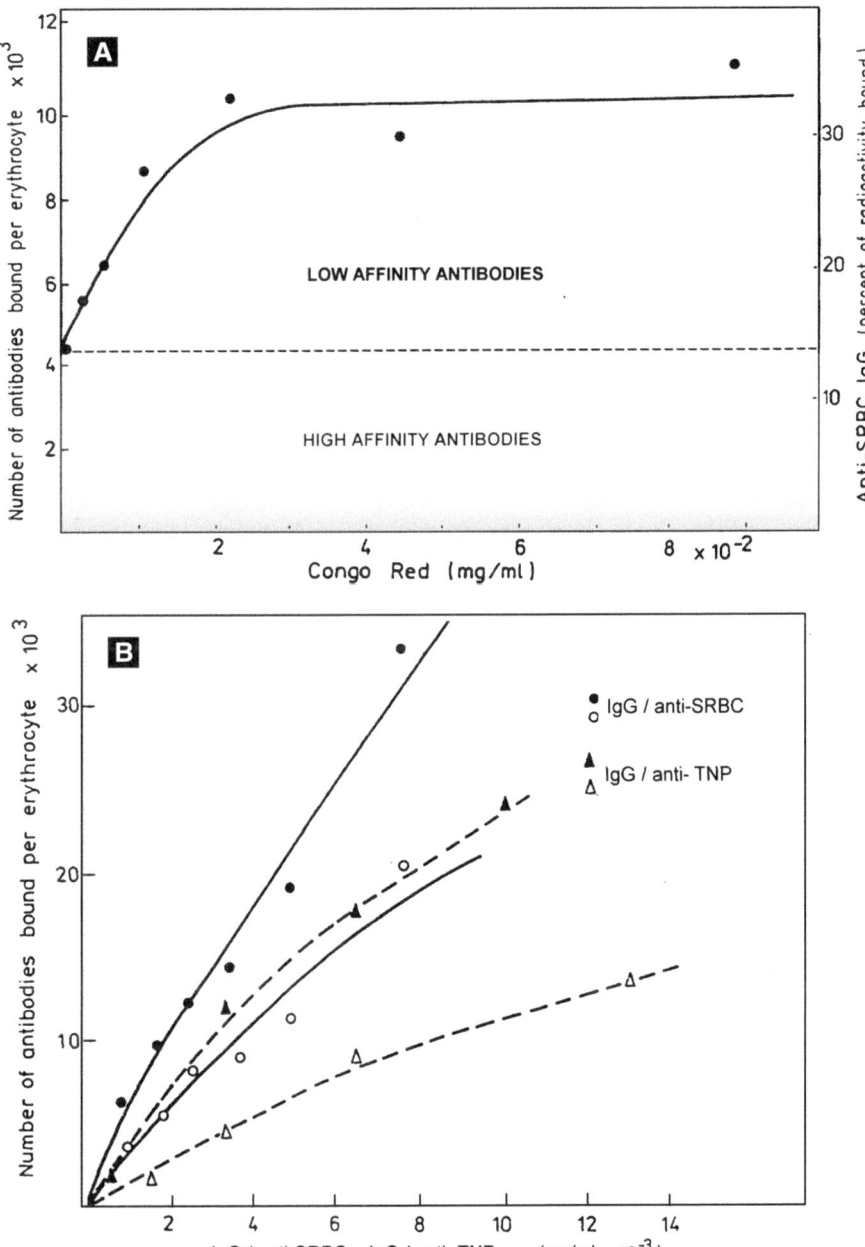

Fig. 2.7 Profiles presenting: (**A**) Increase in the quantity of erythrocyte-agglutinating antibodies under increasing concentrations of CR. High-affinity antibodies are capable of agglutinating cells in the absence of the dye. Anti-SRBC antibodies complex tagged by [125]I. High affinity antibodies – fraction non removable by washing. *Shadowed area* – fraction of highest affinity. (**B**) Enhancement of antibody/antigen interaction by CR, proving that the effect is independent of the type of antigen and specificity of the antibody (Reproduction by permission – *Archivum Immunologiae and Therapie Experimentalis*)

where irreversibility is finally expected – immune complexation, C1q binding etc. On the other hand, the same phenomenon would tend to inhibit the action of enzymes where the ligand must be released following catalysis. Indeed, such inhibition has been confirmed in the scope of complement activation which depends on the action of convertases [33].

The observed enhancement of antibody/antigen complexation capabilities may also be due to another factor: increased flexibility of the V domain, caused by penetration of a large noncovalently stabilized ligand to packing cavity of the replaced N-terminal fragment, bestowing greater internal mobility upon the domain (particularly its CDR loops) and therefore enabling them to align themselves to the antigen with greater accuracy [34].

The discovery and subsequent theoretical study of antigen complexation enhancement triggered by CR creates new possibilities with regard to analysis of intramolecular signal leading to complement system activation – assuming that such signal exists. Due to inhibition of convertase (and therefore of the complement system) by CR, measured as the efficiency of hemolysis, the signal transfer stage (immune complex/C1q) has been separated from the remainder of the activation cascade, including convertase. This reveals activation potential, since both the immune complex and the subsequent complex with C1q are insoluble and may therefore be separated from excess CR by washing, then combined with the remaining components of the complement system, thus preventing undesirable inhibition. To this end we have employed a commercial-grade C1q reagent (QUIDEL USA) and, separately, a serum containing complement system components but deprived of C1q (QUIDEL USA).

Once the excess dye has been washed out, the remaining insoluble complexes (immune complex and immune-C1q complex) prove capable of activating the complement system, triggering hemolysis in C1q-deprived serum. Figure 2.8 presents the results of this experiment. Confirmation of complement system activation reveals the role of CR in the process and confirms the presence of intramolecular signaling.

The immune complex (agglutinate) binds antibodies with varying affinity for red cells which participate in the immune response (SRBC/anti-SRBC). Weak (low-affinity) antibodies are quickly washed out in the absence of CR. The remaining complexes contain antibodies with strong or moderate affinity. The introduction of CR stabilizes the immune complexes formed by weak antibodies, allowing them to remain in the agglutinate. Nevertheless, such antibodies remain incapable of triggering hemolysis even when CR is present. Their properties may be studied by analyzing wash-out samples. Of course, antibodies which resist washing out are also characterized by variable affinity: the group includes strong (high-affinity) antibodies which do not require CR to form stable complexes and trigger hemolysis, but also weak (low-affinity) antibodies stabilized by CR but still unable to trigger hemolysis. The specific affinity threshold established by the wash-out procedure is somewhat arbitrary and depends on a number of conditions. It is assumed that CR account for approximately 50% of the washout-resistant pool. Interestingly, this group contains also antibodies which are unable to trigger complement activation,

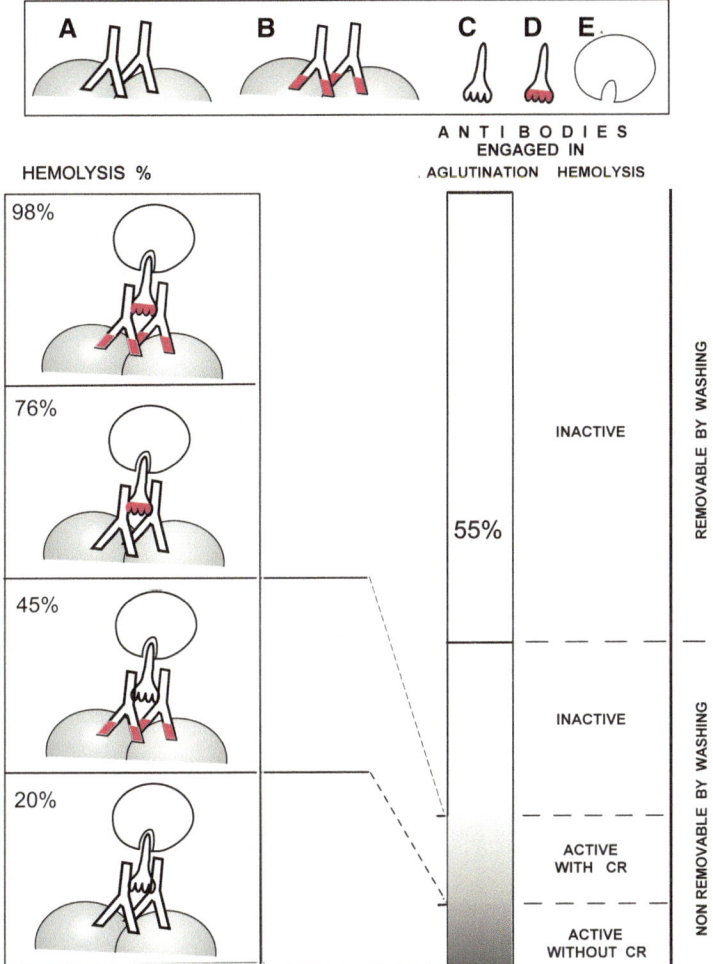

Fig. 2.8 Role of CR in amplifying the complement activation signal. Model view of the process (*left-hand column*). The efficiency of erythrocyte hemolysis by the complement system depends on CR activation of the signaling pathway. Selective action of CR upon successive components of the signal pathway has been marked in red. *A* – immune complex (anti-SRBC antibodies not treated with CR); *B* – antibodies selectively activated by CR; *C* – C1q; *D* – C1q selectively activated by CR; *E* – serum containing all components of the complement system except C1q. Participation of anti-SRBC polyclonal serum antibodies in agglutination and hemolysis is shown on the right

but which gain this ability by interacting with CR. This suggests that a sufficiently powerful intramolecular signal may only be generated by "strong" antibodies which incur significant structural strain when binding the antigen. This natural threshold appears evolutionarily conditioned to prevent accidental thus potentially dangerous activation of the complement system [35–38].

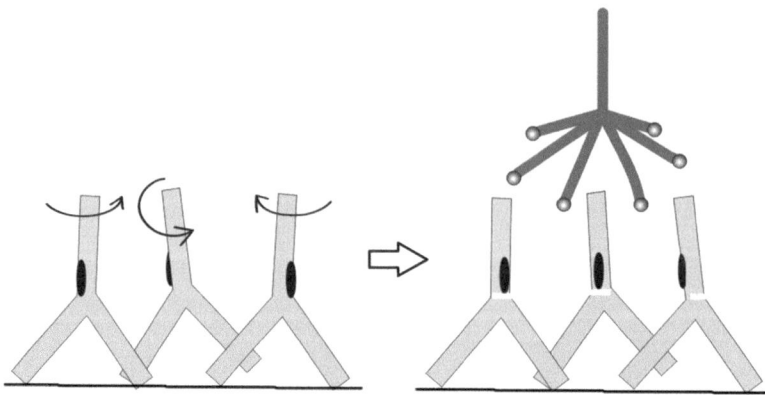

Fig. 2.9 Complexed antibodies showing random placement and orientation of Fc fragments. *Arrows* indicate the need for rotation, facilitating contact with C1q. This is enabled by intramolecular signaling which increases the mobility of Fc fragments

While CR helps explains the signaling mechanism, the origin of the signal itself remains an open question. Since it seems improbable for both V domains (VH and VL) to react to structural strain in the same way, antigen complexation induces torsional stress in the Fab fragment [39]. The resulting rotation uncouples Fc from Fab and allows the former to bind to C1q. This is important, since in the immune complex both the antibodies and their Fc fragments are oriented randomly, which would otherwise hamper complexation of C1q. The uncoupling provides Fc with rotational freedom and enables the link to be established easier (Fig. 2.9).

In this way, the use of a supramolecular dye approaches understanding both the function and the purpose of intramolecular signaling.

2.3   Application of Congo Red for Immunotargeting

The fact that CR selectively binds to antibody/antigen complexes creates an interesting opportunity with regard to targeted drug delivery. CR is not only capable of recognizing the immune complex, but – owing to its supramolecular nature – intercalate various drug particles, acting then as a carrier. Such intercalation should not be regarded as simple mixing of dye and drug molecules – it is more akin to solvation, which involves close interaction between the solvent and the solute. If the solute is structurally flat and presents a planar arrangement of aromatic rings, it can be easily intercalated into the CR micelle. This effect is further enhanced if the solute is positively charged (Fig. 2.10) [34, 40, 41].

The model presented in this chapter comprises CR and rhodamin B (Fig. 2.10). Interaction between both dyes is evidenced by tracking the release of rhodamine B from a dialysis bag in the absence of CR and in a system where both dyes are combined via intercalation. It is readily evident that rhodamine B – itself a supramolecu-

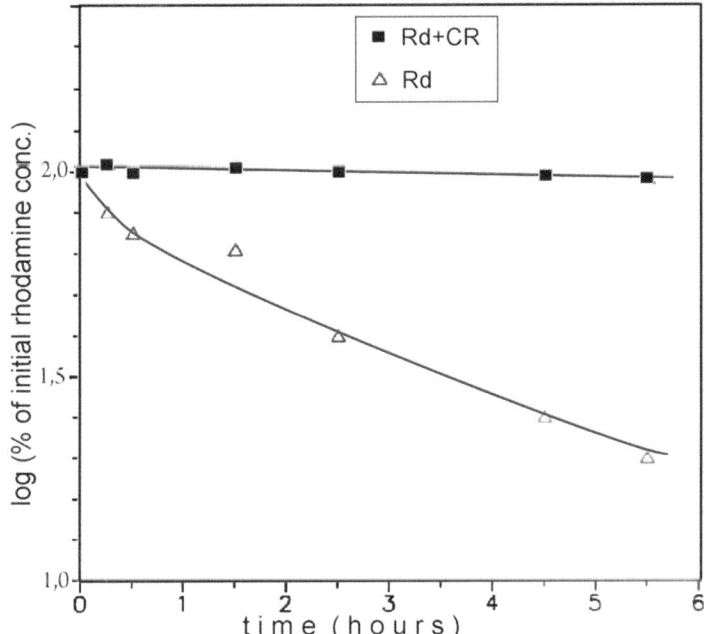

Fig. 2.10 Evidenced by dialysis arresting of rhodamine B (Rd) by supramolecular CR (intercalation), compared to progressive reduction in the concentration of free rhodamine B

lar system, but with weaker self-association tendencies – becomes nearly impervious to dialysis when CR is present. Additional complexes comprising supramolecular dyes and various foreign molecules, analyzed using the dialysis method, are listed in table shown in Fig. 2.11, and are compared to the model CR/rhodamine B complex. Where analysis suggests similar stability of both complexes, a value of 1 is listed. Figure 2.11 presents selected compounds which form co-micellar structures with CR, listing their stability [42]. It seems clear that both the spatial structure and electric charge play an important role.

To further confirm the carrier hypothesis, our analysis focused on a specific immune complex, i.e. agglutination of sheep erythrocytes capable of binding supramolecular ligands. Figure 2.12 presents a visualization of the agglutinate, following addition and subsequent washing out of CR with intercalated rhodamine B [43]. Strong fluorescence of rhodamine B overcomes CR absorption and can be clearly seen along boundaries of agglutinated erythrocytes. This proves that the antibodies involved in agglutination (immune complexes) are bound to the CR/rhodamine B aggregate. Notably, CR does not react with free antibodies – only antibodies engaged in immune complexation can bind the dye. *In vitro* analysis therefore provides evidence of the ability of supramolecular dye to serve as a vehicle for targeted delivery of drugs.

Chemical formula	3-D structure	CS*
Rhodamine B 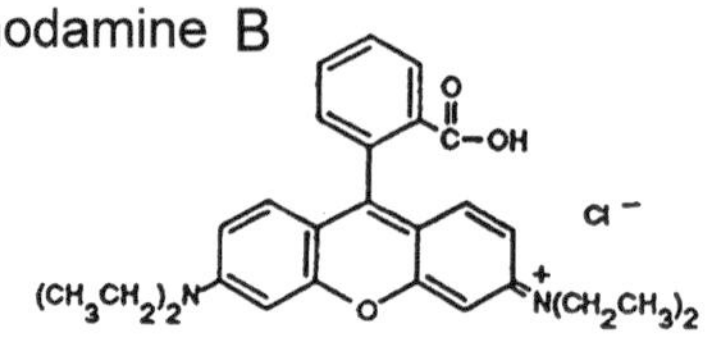	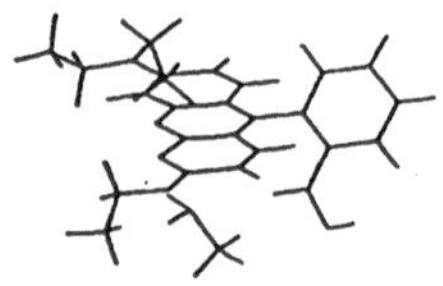	1.00
Rhodamine 6G 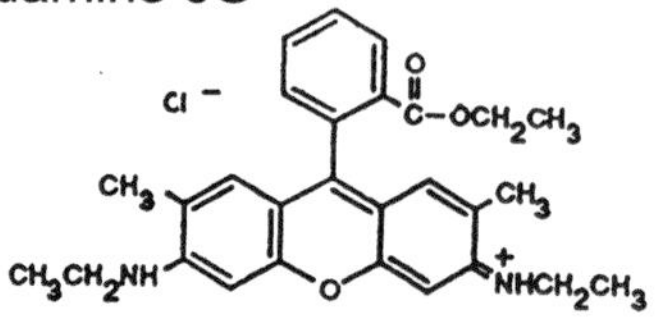	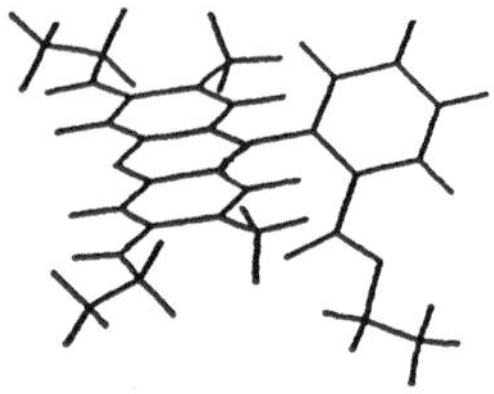	1.00
Adriamycine 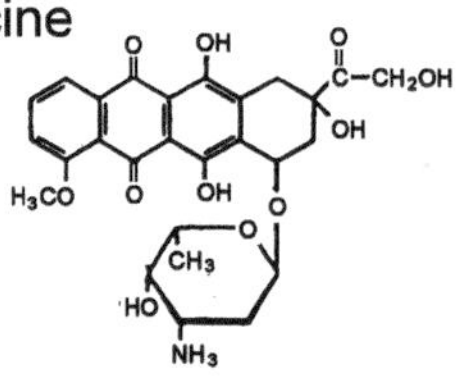	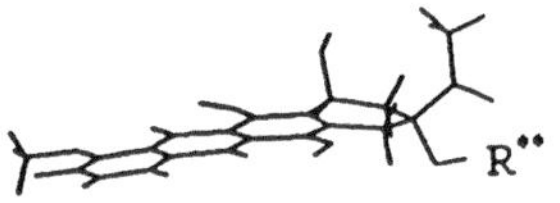	1.00
Methotrexate 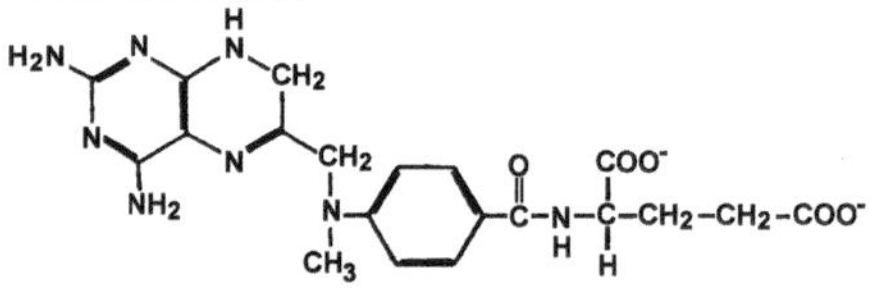	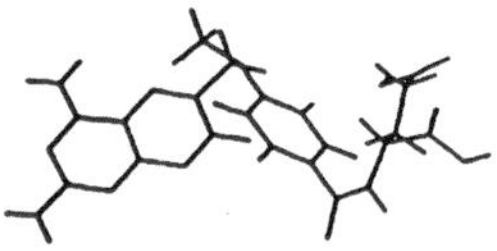	0.09
Fuchsine 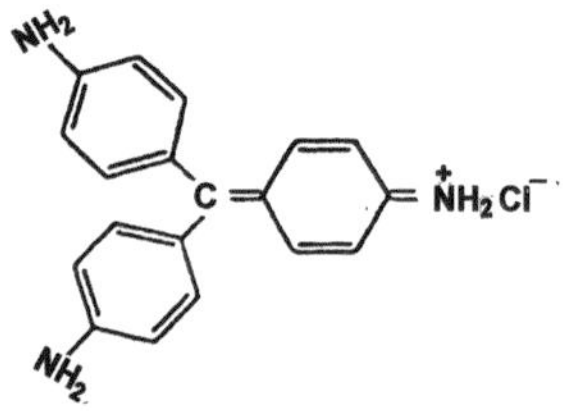	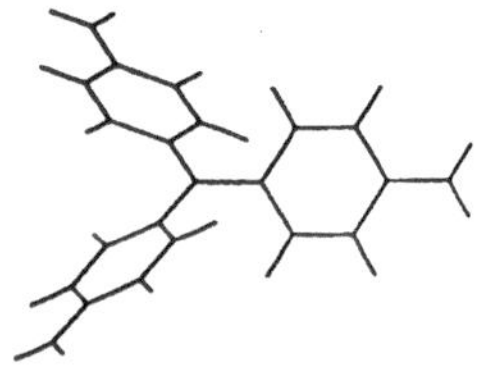	precipitation

Fig. 2.11 Examples of structures readily intercalated by supramolecular CR due to their planar structure and/or positive charge. Complexation tendency ranged 0–1. (Reproduction by permission – *J Physiology and Pharmacology*)

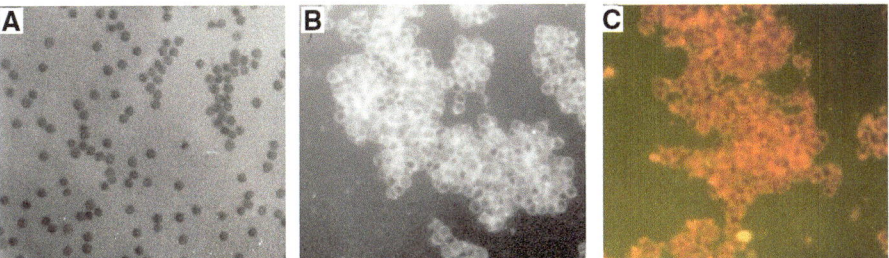

Fig. 2.12 Agglutination in the SRBC/anti-SRBC system produced by antibodies in complex with CR and rhodamine B intercalated. (**A**) not agglutinated red cells – control sample; (**B** and **C**) – agglutinated red cells – UV light (Reproduction by permission – *Folia Histochemica et Cytobiologica*)

It should be noted that *in vitro* test results do not always translate into similar outcomes *in vivo*. Successful use of supramolecular drug carriers in living organisms remains a complex problem due to undesirable reactions with serum proteins, particularly albumin. To determine whether the presented method is feasible, an Arthus reaction has been triggered in a rabbit host – i.e. the TNP antigen (ghosts of rabbit red blood cells conjugated with human IgG) was injected into the earlobe of an sensitized rabbit, producing local inflammation caused by aggregation of immune complexes. The other ear was subsequently injected with 2.5 ml 5 mg/ml) isotonic steryl CR solution. The ear where the Arthus reaction had originally been triggered was then backlit for photographic documentation (the rabbit's thin earlobe is easily penetrated by visible light, simplifying the process). Unfortunately, the unaided human eye is unable to distinguish between CR and hemoglobin – the blood present in the vessels in earlobe produces an color image similar to CR. Enhanced visibility of small blood vessels suggests that an inflammatory process is ongoing, which further hampers attempts to visualize dye accumulations. Effective analysis therefore required the use of specially prepared spectroscopic filters (the spectra of CR and hemoglobin differ somewhat).

Figure 2.13 presents the backlit tissue fragment with spectroscopic filters applied. The modified color scale enables us to easily distinguish the dye and hemoglobin (filter spectra are also presented, showing which wavelengths have been blocked). As expected, the dye is attracted to the antigen injection site, where the immune complex can be found. The experiment also highlights the kinetic characteristics of CR absorption and subsequent removal, showing how the proposed transport system may function in practice (Fig. 2.14).

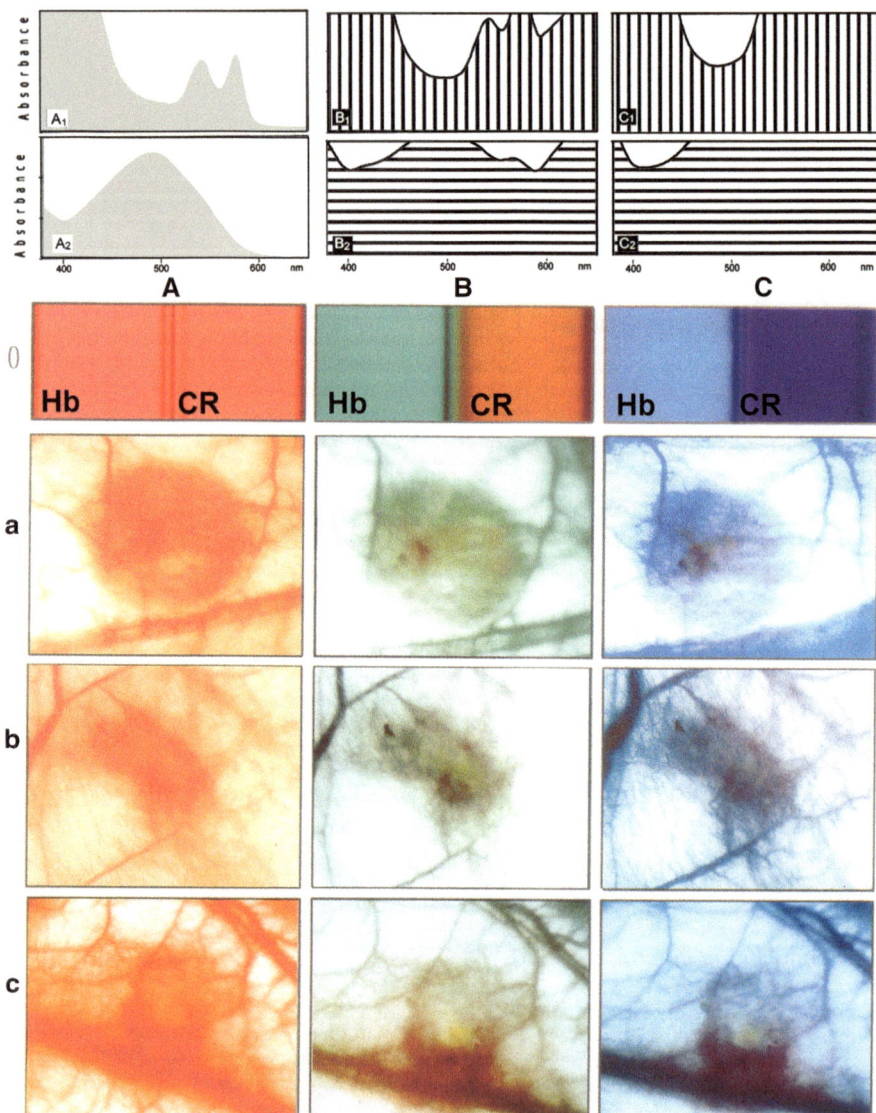

Fig. 2.13 Migration of CR to areas of immune complexation (Arthus reaction induced in rabbite-arlobe). Spectroscopic filters depicted above each column were used to differentiate CR and hemo-globin. Column *A* – spectra of hemoglobin and CR, column *B* and *C* – spectra with filters used (4–6) [43] a, b, c - three independent experiments. (Reproduced by permission – *Folia Histochemica et Cytobiologica*)

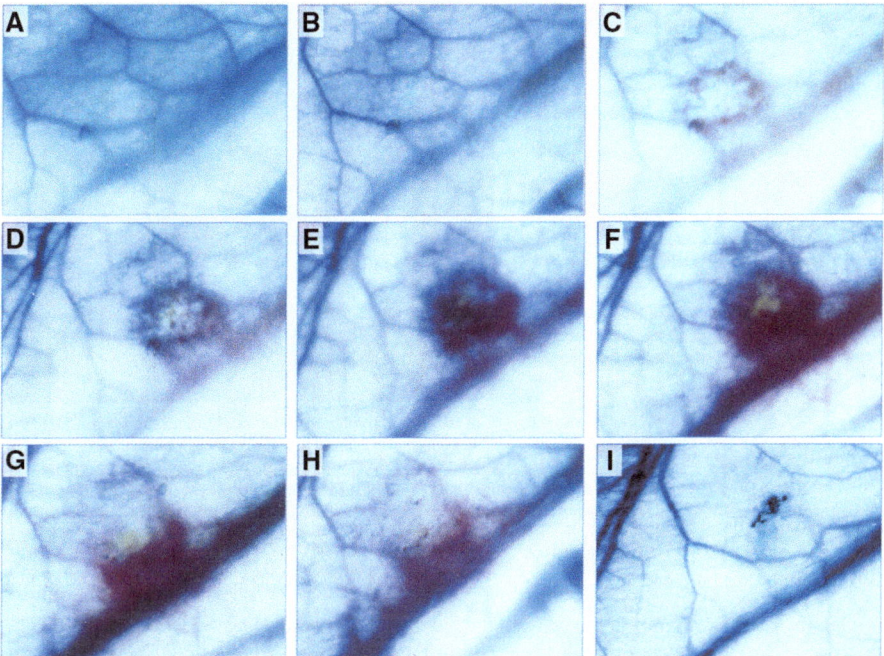

Fig. 2.14 CR accumulation kinetics in the Arthus reaction area: **A** – 1 h35′; **B** – 3 h30′; **C** – 4 h35′; **D** – 5 h; **E** – 9 h30′; **F** – 26 h30′; **G** – 52 h35′; **H** – 72 h20′; **I** – 14 days (Reproduced by permission – *Folia Histochemica et Cytobiologica*)

2.4   Toxicity of CR and Its Applicability for Immunotargeting

An important practical advantage of CR, as well as of other structurally similar supramolecular systems, is their relatively low toxicity. Supramolecular dye aggregations do not readily penetrate cellular membranes, and are easily excreted, along with the surplus of any intercalated substance (Fig. 2.15). Nevertheless, intestinal excretion of CR has been linked to carcinogenesis – most likely due to bacterial reduction of the dye, producing benzidine (a known carcinogen). This undesirable effect may be mitigated by administering a cellulose-rich diet, since cellulose eagerly binds CR and protects it from structural changes, including reduction.

In addition to its immune complexation potential, CR is also being studied in the context of amyloidaffinity, although the presented applications of supramolecular

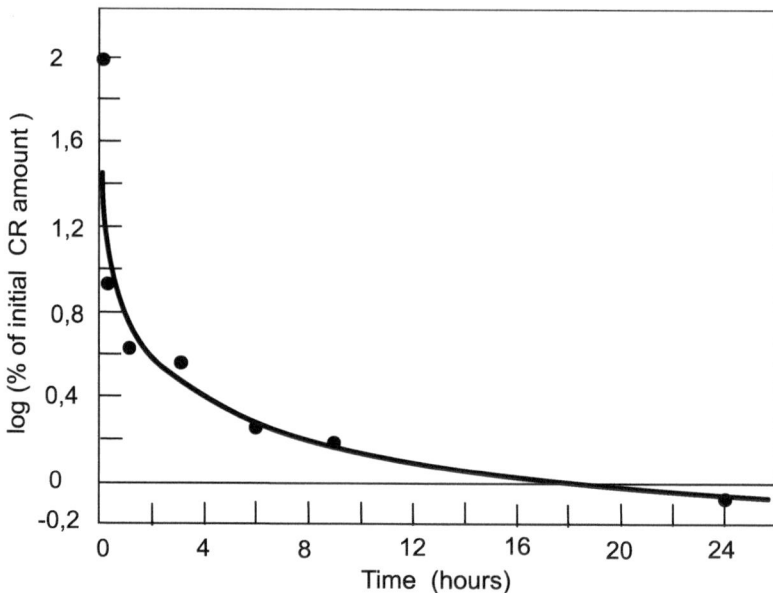

Fig. 2.15 Kinetics of CR clearance from the blood after its intravenous injection (2.5 ml of CR – 5 mg/ml) (Reproduction by permission – *Folia Histochemica et Cytobiologica*)

ligands are based on laboratory experiments. Practical medical applications would require further, independent research.

Acknowledgements We acknowledge the financial support from the National Science Centre, Poland (grant no. 2016/21/D/NZ1/02763) and from the project Interdisciplinary PhD Studies "Molecular sciences for medicine" (co-financed by the European Social Fund within the Human Capital Operational Programme) and Ministry of Science and Higher Education (grant no. K/DSC/001370).

References

1. Edelman GM (1970) The covalent structure of a human gamma G-immunoglobulin. XI Functional implications. Biochemistry 9(16):3197–3205
2. Galvanico NJ, Tomasi TB Jr (1979) Effector sites on antibodies. In: Atassi MZ (ed) Immuno-chemistry of proteins, vol 3. Plenum Press, New York and London, pp 1–85
3. Brekke OH, Michaelsen TE, Sandlie I (1995) The structural requirements for complement activation by IgG: does it hinge on the hinge? Immunol Today 16(2):85–90
4. Harris LJ, Larson SB, McPherson A (1999) Comparison of intact antibody structures and the implications for effector function. Adv Immunol 72:191–208

5. Coloma MJ, Trinh KR, Wims LA, Morrison SL (1997) The hinge as a spacer contributes to covalent assembly and is required for function of IgG. J Immunol 158(2):733–740
6. Cathou RE (1978) Solution conformation an segmental flexibility of immunoglobulins. In: Litman GW, Good RA (eds) Immunoglobulins, vol 5. Plenum Medical Book Company. Comprehensive Immunology, New York/London, pp 37–83
7. Kohoe JM (1978) The structural basis for the biological properties of immunoglobulins. In: Litman GW, Good RA (eds) Immunoglobulins, vol 5. Plenum Medical Book Company, New York/London, pp 173–196
8. Metzger H (1990) General aspects of antibody structure and function. In: Metzger H (ed) Fc receptors and the action of antibodies. American Society Microbiology, Washington, DC, pp 7–11
9. Colman PM, Laver WG, Varghese JN, Baker AT, Tulloch PA, Air GM, Webster RG (1987) Three-dimensional structure of a complex of antibody with influenza virus neuraminidase. Nature 326(6111):358–363
10. Bhat TN, Bentley GA, Fischmann TO, Boulot G, Poljak RJ (1990) Small rearrangements in structures of Fv and Fab fragments of antibody D1.3 on antigen binding. Nature 347(6292):483–485
11. Schulze-Gahmen U, Rini JM, Wilson IA (1993) Detailed analysis of the free and bound conformations of an antibody. X-ray structures of Fab 17/9 and three different Fab-peptide complexes. J Mol Biol 234(4):1098–1118
12. Schneider WP, Wensel TG, Stryer L, Oi VT (1988) Genetically engineered immunoglobulins reveal structural features controlling segmental flexibility. Proc Natl Acad Sci U S A 85(8):2509–2513
13. Sensel MG, Kane LM, Morrison SL (1997) Amino acid differences in the N-terminus of C(H)2 influence the relative abilities of IgG2 and IgG3 to activate complement. Mol Immunol 34(14):1019–1029
14. Sandlie I, Michaelsen TE (1991) Engineering monoclonal antibodies to determine the structural requirements for complement activation and complement mediated lysis. Mol Immunol 28(12):1361–1368
15. Brekke OH, Michaelsen TE, Sandin R, Sandlie I (1993) Activation of complement by an IgG molecule without a genetic hinge. Nature 363(6430):628–630
16. Rybarska J, Konieczny L, Bobrzecka K, Laidler P (1982) The hemolytic activity of (Fab-Fc) recombinant immunoglobulins with specificity for the sheep red blood cells. Immunol Lett 4(5):279–284
17. Rybarska J, Konieczny L, Roterman I, Piekarska B (1991) The effect of Azo dyes on the formation of immune complexes. Arch Immunol Ther Exp (Warsz) 39:317–327
18. Stopa B, Górny M, Konieczny L, Piekarska B, Rybarska J, Skowronek M et al (1998) Supramolecular ligands: monomer structure and protein ligation capability. Biochimie 80:963–968
19. Kaszuba J, Konieczny L, Piekarska B, Roterman I, Rybarska J (1993) Bis-azo dyes interferencje with effector activation of antibodies. J Physiol Pharmacol 44:233–242
20. Glenner GG, Eanes ED, Page DL (1972) The relation of the properties of Congo red-stained amyloid fibrils to the -conformation. J Histochem Cytochem 20(10):821–826
21. Skowronek M, Stopa B, Konieczny L, Rybarska J, Spólnik P, Zemanek G et al (2003) The structure and protein-binding of amyloid-specific dye reagents. Acta Biochim Pol 50:1213–1227
22. Stopa B, Piekarska B, Konieczny L, Rybarska J, Spólnik P, Zemanek G, Roterman I, Król M (2003) The structure and protein binding of amyloid-specific dye reagents. Acta Biochim Pol 50(4):1213–1227
23. Rybarska J, Piekarska B, Stopa B, Zemanek G, Konieczny L, Nowak M, Król M, Roterman I, Szymczakiewicz-Multanowska A (2001) Evidence that supramolecular Congo red is the

sole ligation form of this dye for L chain lambda derived amyloid proteins. Folia Histochem Cytobiol 39(4):307–314
24. Roterman I, No KT, Piekarska B, Kaszuba J, Pawlicki R, Rybarska J, Konieczny L (1993) Bis azo dyes – studies on the mechanism of complex formation with IgG modulated by heating or antigen binding. J Physiol Pharmacol 44(3):213–232
25. Roterman I, Rybarska J, Konieczny L, Skowronek M, Stopa B, Piekarska B et al (1998) Congo red bound to α-1 proteinase inhibitor as a model of supramolecular ligand and protein complex. Comput Chem 22:61–70
26. Piekarska B, Konieczny L, Rybarska J, Stopa B, Zemanek G, Szneler E, Król M, Nowak M, Roterman I (2001) Heat-induced formation of a specific binding site for self-assembled Congo red in the V domain of immunoglobulin L chain lambda. Biopolymers 59(6):446–456
27. Ewert S, Honegger A, Plückthun A (2004) Stability improvement of antibodies for extracellular and intracellular applications: CDR grafting to stable frameworks and structure-based framework engineering. Methods 34(2):184–199
28. Röthlisberger D, Honegger A, Plückthun A (2005) Domain interactions in the Fab fragment: a comparative evaluation of the single-chain Fv and Fab format engineered with variable domains of different stability. J Mol Biol 347(4):773–789
29. Smith TJ, Olson NH, Cheng RH, Chase ES, Baker TS (1993) Structure of a human rhinovirus-bivalently bound antibody complex: implications for viral neutralization and antibody flexibility. Proc Natl Acad Sci U S A 90(15):7015–7018
30. Thouvenin E, Laurent S, Madelaine MF, Rasschaert D, Vautherot JF, Hewat EA (1997) Bivalent binding of a neutralising antibody to a calicivirus involves the torsional flexibility of the antibody hinge. J Mol Biol 270(2):238–246
31. Pilz I, Kratky O, Licht A, Sela M (1975) Shape and volume of fragments Fab' and (Fab')2 of anti-poly(D-alanyl) antibodies in the presence and absence of tetra-D-alanine as determined by small-angle x-ray scattering. Biochemistry 14(6):1326–1333
32. Piekarska B, Konieczny L, Rybarska J, Stopa B, Spólnik P, Roterman I, Król M (2004) Intramolecular signaling In immunoglobulins – new evidence emerging from the use of supramoelcular proteins ligands. J Physiol Pharmacol 55:487–501
33. Krol M, Roterman I, Drozd A, Konieczny L, Piekarska B, Rybarska J, Spolnik P, Stopa B (2006) The increased flexibility of CDR loops generated in antibodies by Congo red complexation favors antigen binding. J Biomol Struct Dyn 23(4):407–416
34. Jagusiak A, Konieczny L, Król M, Marszałek P, Piekarska B, Piwowar P, Roterman I, Rybarska J, Stopa B, Zemanek G (2015) Intramolecular immunological signal – hypothesis reviewed – structural background of signalling revealed by Rusing Congo red as a specific tool. Mini Rev Med Chem 14(13):1104–1113
35. Hamada A, Watanabe N, Azuma T, Kobayashi A (1990) Enhancing effect of C1q on IgG monoclonal antibody binding to hapten. Int Arch Allergy Appl Immunol 91(1):103–107
36. Tan LK, Shopes RJ, Oi VT, Morrison SL (1990) Influence of the hinge region on complement activation, C1q binding, and segmental flexibility in chimeric human immunoglobulins. Proc Natl Acad Sci U S A 87(1):162–166. Erratum Proc Natl Acad Sci USA 1991 88, 5066
37. Gaboriaud C, Juanhuix J, Gruez A, Lacroix M, Darnault C, Pignol D, Verger D, Fontecilla-Camps JC, Arlaud GJ (2003) The crystal structure of the globular head of complement protein C1q provides a basis for its versatile recognition properties. J Biol Chem 278(47):46974–46982
38. Vargas-Madrazo E, Paz-Garcia E (2003) An improved model association for VH-VL immunoglobulin domains: asymmetries between VH and VL in the packing of some interface residues. J Mol Recognit 16:113–120

39. Skowronek M, Stopa B, Konieczny L, Rybarska J, Piekarska B, Szneler E, Bakalarski G, Roterman I (1998) Self-assembly of Congo red – a theoretical approach to identify its supramolecular organization In water and salt solutions. Biopolymers 46:267–281
40. Król M, Roterman I, Piekarska B, Konieczny L, Rybarska J, Stopa B, Spólnik P, Szneler E (2005) An approach to understand the complexation of supramolecular dye Congo red with immunoglobulin L chain lambda. Biopolymers 77(3):155–162
41. Konieczny L, Piekarska B, Rybarska J, Stopa B, Krzykwa B, Noworolski J, Pawlicki R, Roterman I (1994) Bis azo dye liquid crystalline micelles as possible drug carriers in immunotargeting technique. J Physiol Pharmacol 45(3):441–454
42. Konieczny L, Piekarska B, Rybarska J, Skowronek M, Stopa B, Tabor B, Dabroś W, Pawlicki R, Roterman I (1997) The use of congo red as a lyotropic liquid crystal to carry stains in a model immunotargeting system-microscopic studies. Folia Histochem Cytobiol 35(4):203–210
43. Rybarska J, Piekarska B, Stopa B, Spólnik P, Zemanek G, Konieczny L, Roterman I (2004) In vivo accumulation of self-assembling dye Congo red in an area marked by specific immune complexes: possible relevance to chemotherapy. Folia Histochem Cytobiol 42(2):101–110

Chapter 3
Protein Conditioning for Binding Congo Red and Other Supramolecular Ligands

Grzegorz Zemanek, Anna Jagusiak, Joanna Rybarska, Piotr Piwowar, Katarzyna Chłopaś, and Irena Roterman

Abstract Self-assembled organic compounds which form ribbon-like micellar clusters may attach themselves to proteins, penetrating in areas of low stability. Such complexation involves regions other than the protein's natural binding site. The supramolecular ligand adheres to beta folds or random coils which become susceptible to complexation as a result of function-related structural changes – e.g. antibodies engaged in immune complexes or acute phase proteins. However, even seemingly unsusceptible helical proteins may bind Congo red if they include chameleon sequences (short peptide fragments capable of adopting different secondary conformations depending on environmental conditions). Examples of such proteins include hemoglobin and albumin. Complexation of supramolecular Congo red is often associated with increased fluorescence, indicating breakdown of ligand micelles in the complex. This phenomenon may be used in diagnostic tests.

Keywords Congo red binding to haemoglobin • Acute phase • Congo red binding to albumin • Fluorescence of Congo red • Cell-cell interaction and Congo red binding • Proteins susceptibility for protein complexation • Chameleon sequences and Congo red binding

G. Zemanek (✉) • A. Jagusiak • J. Rybarska • K. Chłopaś
Chair of Medical Biochemistry, Jagiellonian University – Medical College,
Kopernika 7, 31-034 Krakow, Poland
e-mail: grzegorz.zemanek@uj.edu.pl; anna.jagusiak@uj.edu.pl; mbstylin@cyf-kr.edu.pl;
katarzyna.chlopas@wp.pl

P. Piwowar
Department of Measurements and Electronics, AGH University of Science and Technology,
Adama Mickiewicza 30, 30-059, Krakow, Poland
e-mail: ppiwowar@agh.edu.pl

I. Roterman
Department of Bioinformatics and Telemedicine, Jagiellonian University – Medical College,
Łazarza 16, 31-530, Krakow, Poland
e-mail: myroterm@cyf-kr.edu.pl

© The Author(s) 2018
I. Roterman, L. Konieczny (eds.), *Self-Assembled Molecules – New Kind of Protein Ligands*, https://doi.org/10.1007/978-3-319-65639-7_3

3.1 Natural Susceptibility of Proteins to Bind Congo Red

As remarked in Chaps. 1 and 2, supramolecular micellar systems may form complexes with proteins by penetrating in areas other than the active site. This process is conditioned by local instabilities in the protein structure [1, 2].

As a rule not all parts of a protein molecule are equally stable. Local instabilities are usually not significant enough to enable direct penetration of a large ligand consisting of many associated molecules. Such instabilities can, however, be artificially exacerbated, e.g. through heating [3]. Under natural conditions Congo red (CR) is spontaneously bound by partly unfolded proteins capable of forming aggregations – such as amyloids and some abnormal [4–9]. While aberrant (unstable) proteins are usually eliminated before they can leave their parent cell [10–13], under certain circumstances – such as mass synthesis of light chains associated with multiple myeloma – they can be detected in circulation (Fig. 3.1) [14–19].

On the other hand, proteins destabilized partly through complexation of their "intended" natural ligands are commonly found in bodily fluids, especially in blood serum. Such proteins usually acquire the ability to bind CR. Examples include acute-phase proteins, conditioned to capture and eliminate other proteins whose presence in the bloodstream is harmful – e.g. proteolytic enzymes (which – due to their specific mechanism of action – may penetrate from the digestive tract or be released by necrotic cells), as well as hemoglobin, which has undesirable catalytic

Fig. 3.1 Standard electrophoretic separation of serum proteins with revealed monoclonal fractions (thick bands in columns 9–12 – mostly L chains)

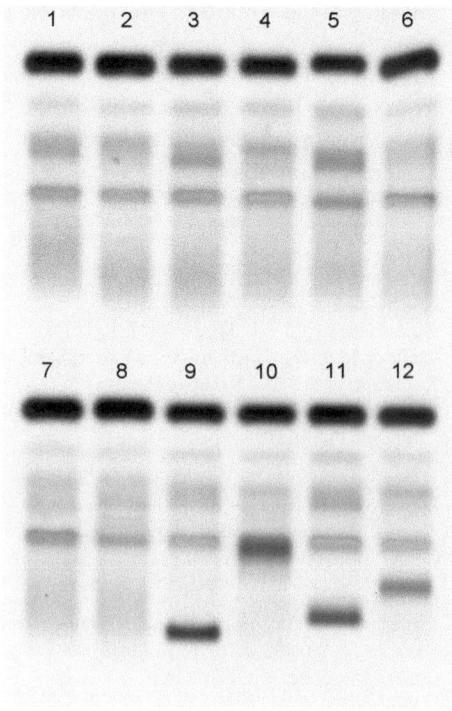

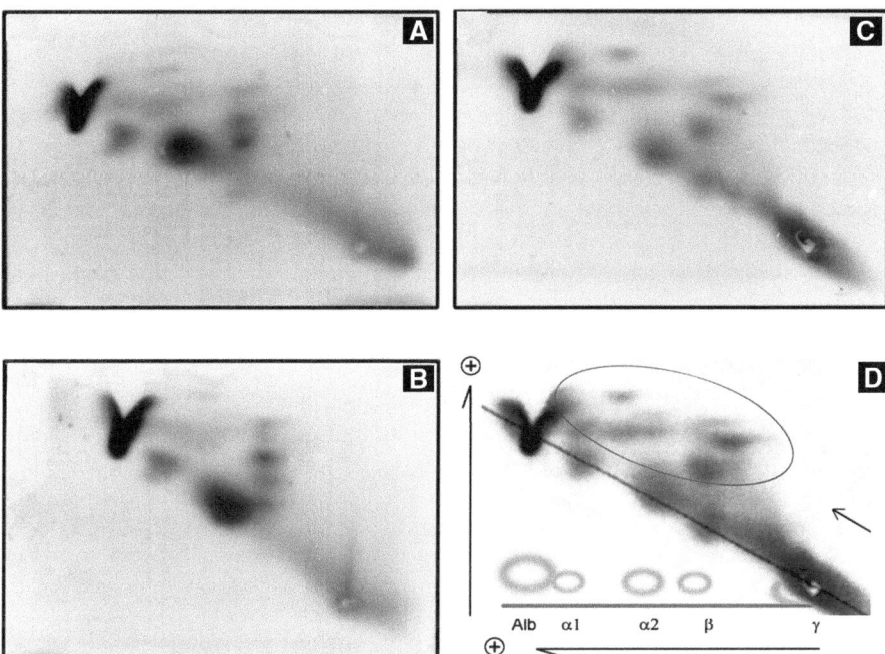

Fig. 3.2 Two-directional agarose electrophoresis with exposed CR binding fractions dislocated over the diagonal line which includes non-binding serum proteins (**A**, **B**, **C** - three independent examples). (**D**) Presents the basis for this method. The *red band* indicates CR, which is spread on the agarose plate before the second electrophoretic step

properties and should not be present in the bloodstream in its unbound form. These acute phase complexes are similar to antibody/ligand complexes – by binding its natural ligand the protein undergoes structural deformations which facilitate penetration of an additional supramolecular ligand [20]. Acute phase complexes are short-lived, since they are recognized and removed by liver enzymes and phagocytes. Nevertheless, as long as the underlying cause persists, the blood serum will always contain proteins capable of association with CR. The specific composition of acute phase proteins present depend on the ongoing pathological processes. Acute phase proteins include serpins, haptoglobin, ceruloplasmin, ferritin, complement factors C3 and C4, as well as albumin, prealbumin and transferrin [21] (although note that in the last three cases pathological processes do not increase the concentration of the corresponding protein, but decrease it instead).

The ability of CR to bind to serum proteins is readily evidenced by agarose gel electrophoresis where the supramolecular ligand is added to the column at a certain stage of the process, accelerating migration of the affected proteins. Figure 3.2 presents a typical scenario where serum proteins are subjected to two-stage electrophoresis on agarose gel, with the second stage carried out in a perpendicular direction to the first, in the presence of CR (red band in Fig. 3.2D) [3, 21].

The dye quickly migrates towards the anode and bypasses most proteins, except those structurally conditioned to bind the dye. As a result, migration of the affected proteins is accelerated and those proteins are found above the diagonal line which collects non bonding proteins. The mobility imparted by CR upon the target protein depends on the number of associated dye molecules and the molecular weight of the protein itself – hence the variable results observed under electrophoresis. The presented technique could have important diagnostic applications; here, however, our goal is merely to present the interaction of CR with serum proteins in order to explain the mechanism of supramolecular complexation. Regarding acute phase proteins, CR complexation is particularly evident in the case of proteins which always exhibit some form of activity, but whose activity in pathological conditions is significantly increased and/or altered.

Some proteins can bind supramolecular ligands even in the absence of a complexation partner which would account for structural rearrangement and reduction in stability. This includes albumin. Owing to its function, albumin is capable of binding various anionic compounds, of which the dye is an example. CR is furthermore capable of associating with amyloidogenic apolipoproteins [22–26].

3.2   Interaction of Congo Red with Helical Structures of Polypeptides

With regard to CR complexation capabilities, haptoglobin represents another interesting research target. The protein binds free hemoglobin, dissociating it into its alpha/beta subunits, but persists as a bridge between both halves [27]. The resulting structure can form complexes with CR. The dye itself also causes dissociation of hemoglobin into identical alpha/beta subunits, but releases them without forming a bridge. Hemoglobin is a typical allosteric protein, with the associated instability most likely concentrated in its alpha/beta interface region. This instability promotes complexation of supramolecular ligands. It appears that CR induces local changes in the alpha-helix structure of the subunit alpha of hemoglobin, transforming it into a beta fold or a random coil, which aids complexation. The process is reversible – adsorption of CR (e.g. on the P15 gel, which strongly binds the dye) yields normal hemoglobin tetramers. Increasing concentrations of the dye produce stronger dissociation – see Fig. 3.3 for electrophoretic images of ½ Hb/CR complexation activity. The effect is independently confirmed by DLS measurements, revealing in the mixture the reaction products of lower molecular weight (Fig. 3.4).

Dissociation of hemoglobin can also be observed when treating hemoglobin crystals (horse) with CR. The resulting dissolution proves that, as a result of binding the dye, the protein undergoes structural changes which prevent crystallization (Fig. 3.5) [28–31].

Probable complexation loci in hemoglobin subunits have been identified by locating structurally unstable alpha folds, which also exhibit measurable propensity

Fig. 3.3 Complexation of dog hemoglobin with CR. Bars representing decreasing amount of dog hemoglobin (slow moving electrophoretic fraction) caused by its transfer to the fast-migrating complex with CR (inset). The CR dye concentrations in Hb were: *1* 1.0×10^{-6} M/ml; *2* 2×10^{-6} M/ml; *3* 4×10^{-6} M/ml; *4* 5×10^{-6} M/ml; *5* 7.5×10^{-6} M/ml

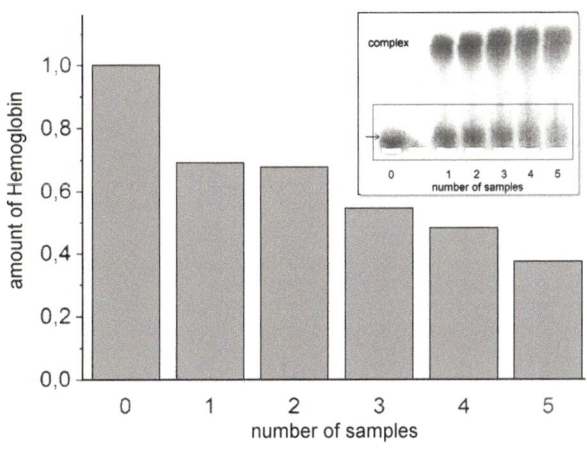

Fig. 3.4 DLS analysis revealing a shift towards lower-mass molecules following interaction of hemoglobin with CR (red bar), confirming dissociation of the protein into subunits

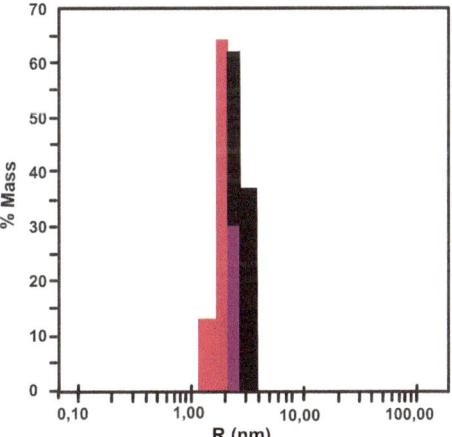

towards adopting beta or random coil conformations. Such folds may favor penetration of the supramolecular ligand, altering their own conformation in the process and producing finally stable bond. Comparative analysis has been performed by querying a database of chameleon sequences (i.e. sequences which may adopt either alpha or beta conformations, depending on local conditions) [32]. For each tetrapeptide, the database lists the corresponding probabilities of encountering alpha, beta and random coil conformations in actual proteins – this enables identification of fragments of the hemoglobin chain which may potentially transit to beta folds or random coil.

Results are presented in Fig. 3.6 as a sequence of bar charts illustrating the structural propensities of each residue sequence.

While chameleon sequences are found in both subunits of the hemoglobin chain, their placement in the alpha subunit appears to be more favorable for binding CR – they are located close to one another and provide a convenient pocket with beta or

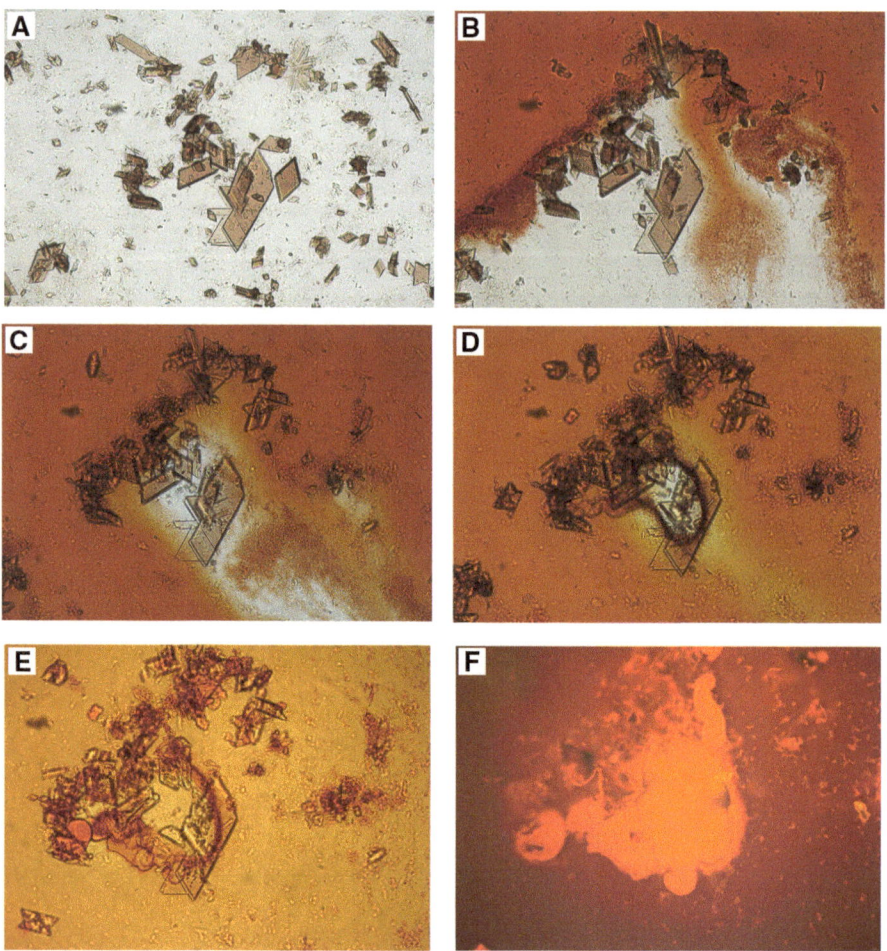

Fig. 3.5 Disintegration of hemoglobin crystals (horse) upon interaction with CR observed under a microscope, in visible light (**A–D**) and UV (**E** and **F**)

random folds found on either side of the ligand. Experimental evidence suggests that CR penetrates the protein as a supramolecular ligand, wedging itself between parallel folds as long as such folds are not tightly packed and may be induced to adopt beta or random coil conformations. In contrast, helical folds do not support complexation of CR due to steric clashes. Regarding the alpha subunit, chameleon fragments are found in areas referred to as the G, H and FG helices, all located in close proximity of the heme. Structural rearrangement caused by complexation of CR therefore causes dissociation of the protein.

Figure 3.7 provides a space-filling depiction of the abovementioned fragments, while Fig. 3.8 illustrates the transition of unstable helices into beta folds or random coils, using albumin as an example. While albumin is ordinarily a helical protein,

Fig. 3.6 Bars representing the tendency of successive polypeptide chain fragments (numbers listed on the horizontal axis) of the human hemoglobin alpha subunit to adopt alpha, beta and random coil conformations respectively. The presented values are derived from the chameleon sequence database [32] (**A**) The same concerning neighbor amino acids (**B**)

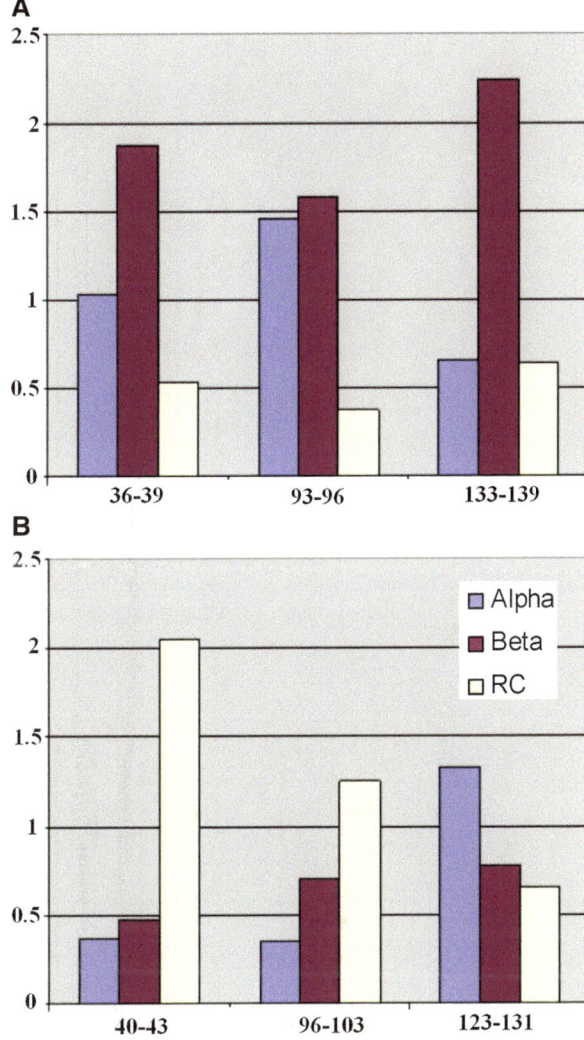

complexation of CR produces a notable conformational shift in favor of beta folds and random coils. Interestingly, similar interaction with EB and TB does not have a similar effect on the secondary conformation of the albumin chain [26] (Fig. 3.8).

It should be noted that the CR micelle is much more cohesive than its EB counterpart and therefore exerts a greater structural influence upon its surroundings. As a result, this relatively stable supramolecular ligand may force the target protein to adapt to its conformational preferences. This contrasts with EB, which relies on natural complexation sites present in the protein and does not induce conformational changes in its alpha chains. As already remarked, such "forcing" action of CR affects chameleon sequences which nominally appear as helices, but may also adopt beta or random coil conformations. Stable alpha folds do not yield to the presence of CR – although the stability threshold beyond which conformational changes

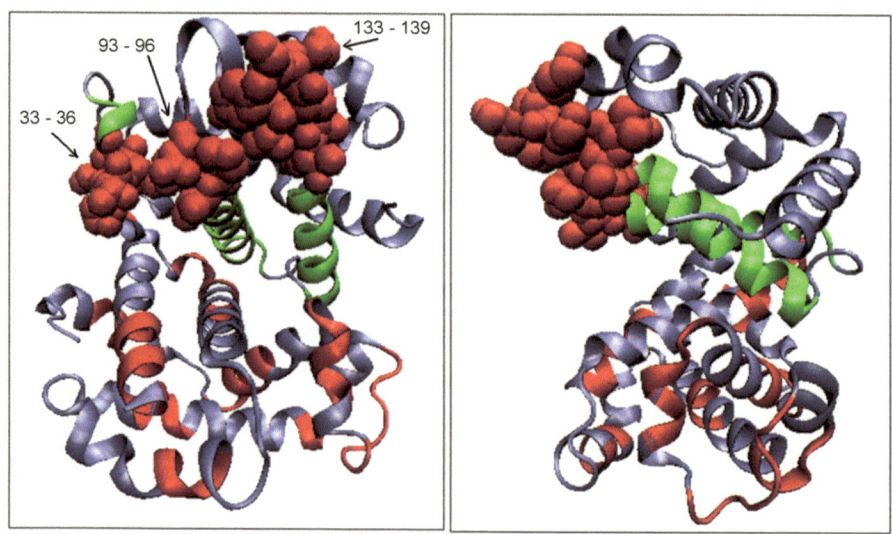

Fig. 3.7 Alpha/beta subunit of hemoglobin (3OO5 – PDB) with chameleon fragments likely to adopt beta and random coil conformations marked in *red*, and additionally presented in a space-filling model in the alpha subunit. The images are rotated by 90° with respect to each other

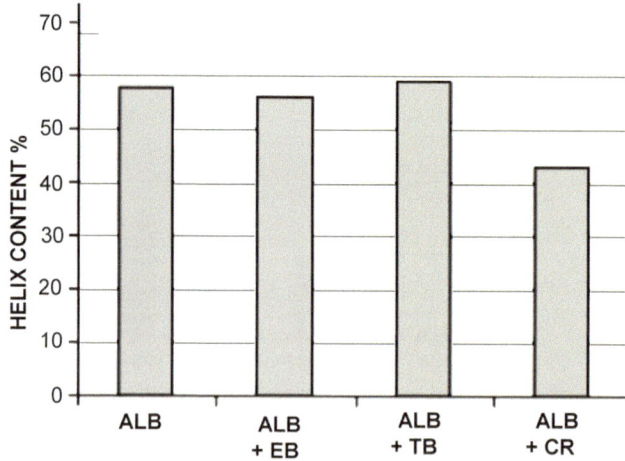

Fig. 3.8 *Bars* representing the decrease in the quantity of alpha folds in albumin upon binding supramolecular ligands – particularly CR, indicating that compactness of the ligand is critical for complexation

occur is not well defined. It appears that increasing the concentration of CR results in more cohesive micelles with greater protein penetration ability. This is suggested by electrophoretic migration rate which increases as the acidity of CR in solution grows, indicating changes in pK of sulfonic group (Fig. 3.9).

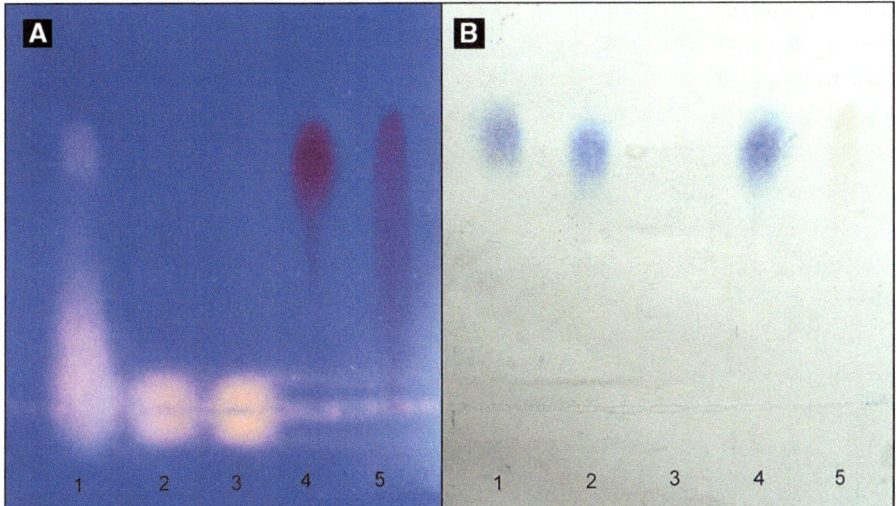

Fig. 3.9 Fluorescent albumin spot seen on an electrophoretic plate, showing that positively charged rhodamine B molecules with no affinity to albumin may be introduced to it anyway via CR which binds rhodamine B by intercalation. Agarose electrophoresis. *1* albumin combined with CR and rhodamine B (complex); *2* albumin + rhodamine B; *3* rhodamine B; *4* albumin + CR; *5* CR. (**A**) UV image. (**B**) Visible light image following reduction and protein staining

Having induced structural changes in the protein, CR forms a complex with the newly produced beta fold or random coil, stabilizing the change. The dye penetration limit is determined by mutual alignment between the protein's secondary folds and the dye micelle. An important factor facilitating penetration is the cohesiveness of the ligand itself, which – unlike the polypeptide – is not a polymer but rather an associate, stabilized by noncovalent interactions [33–35].

Albumin is a typical helical protein with a particularly vital role in blood serum. The importance of albumin is underscored by its natural concentrations and deep involvement in energy management processes. Albumin represents a source of amino acids in cases of malnutrition, but it also serves as a carrier of fatty acids. It is further capable of binding and transporting a variety of dyes and drugs [27]. Its complexation affinity for supramolecular CR and EB [26] makes it a useful study subject. Clearly, albumin may play a role in immunotargeting, since it is capable of forming complexes with supramolecular ligands doped with therapeutic agents.

The binding of supramolecular CR and EB by albumin has been confirmed by synthesizing co-micellar structures, i.e. supramolecular structures consisting of either dye mixed with a foreign compound a positively charged dye not normally complexed by albumin, such as rhodamine B or Janus Green, which can only be bound to the protein as an intercalant. In the case of rhodamine B, this effect is revealed by UV imaging of electrophoretic plates. The characteristic fluorescence of rhodamine B coincides with the location of the albumin stain, proving that the dye enters the protein as a component of a co-micellar structure formed by CR or

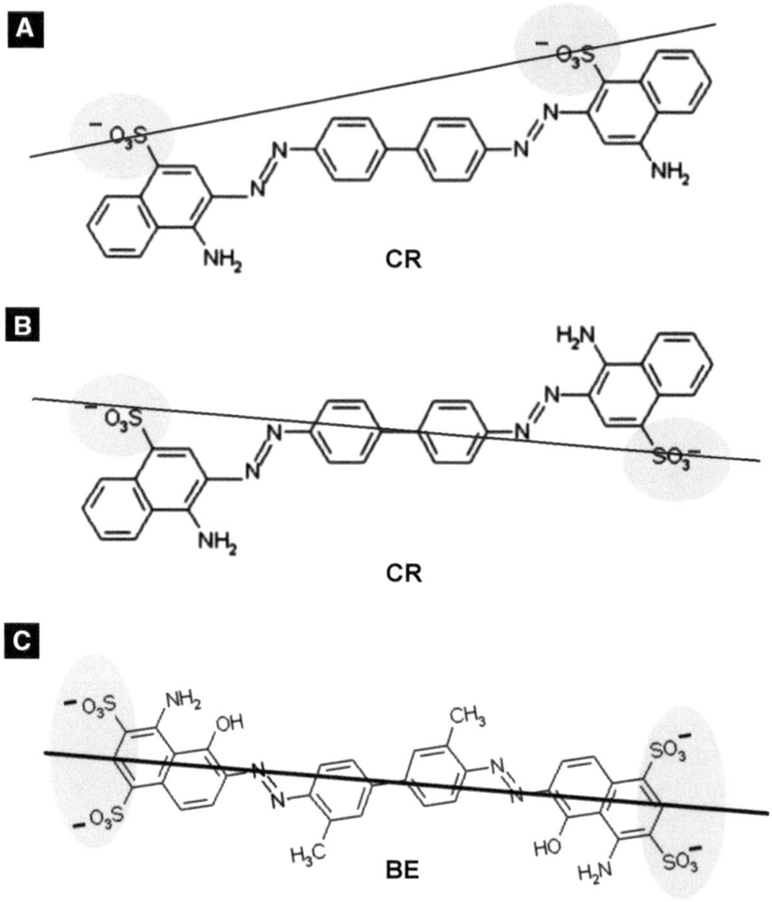

Fig. 3.10 The geometry and charge distribution of CR and EB dyes, explaining the prevalence of the form trans in the case of CR rotamers. (**A**) CR cis, (**B**) CR trans, (**C**) EB

EB (Fig. 3.10). The supramolecular nature of the ligand bound to albumin is also confirmed by counting the number of dye molecules attached to each albumin molecule (16–20 unit molecules in the case of CR) [26]. The link between self-association properties and protein complexation potential further suggests that the ligand functions as a supramolecular entity.

Additional insight into the specifics of CR and EB complexation with albumin is provided by spectro-polarimetric analysis. The complex with EB is strongly chiral, while the corresponding complex with CR does not exhibit chirality [26]. This suggests that EB favors cis binding, while CR binds in a trans (alternating) alignment. The effect can be explained by referring to the structure of both dyes. In EB, all sulfonic groups are adjacent and bound to aromatic rings at either end of the molecule and therefore aligned with its long axis. This results in mutual modification of

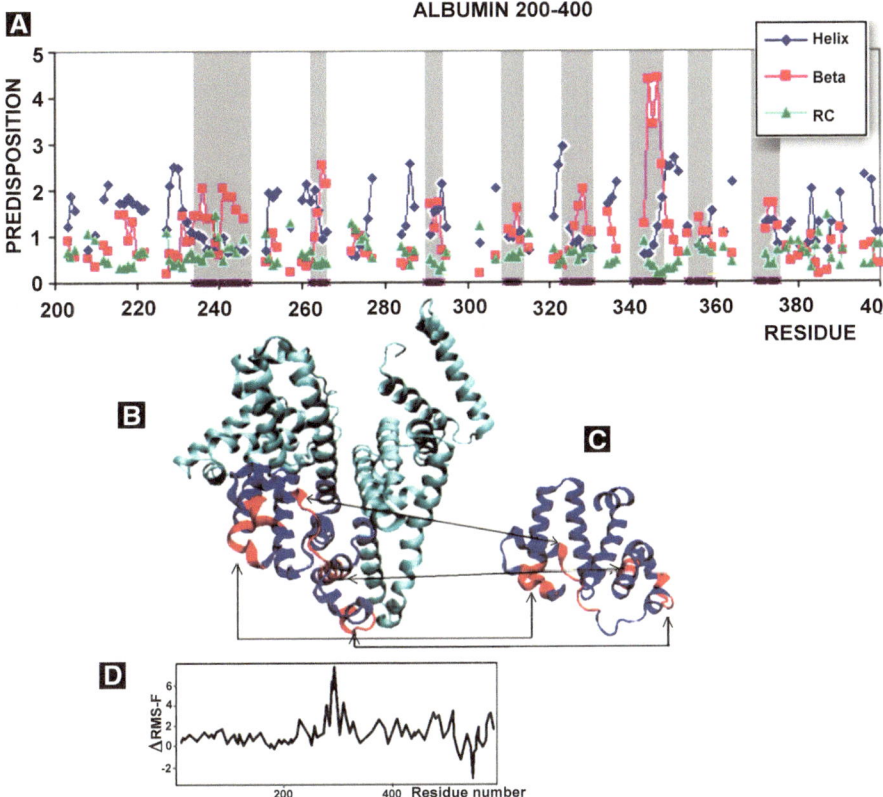

Fig. 3.11 The most unstable area of albumin (200–400 aa) estimated according to secondary structure predisposition. (**A**) Profile indicating the predisposition to adopt helical (H), beta (B) and random coil (RC) conformations of each chain fragment. Fragments particularly likely to adopt beta forms are marked on the horizontal axis (*shaded areas*). (**B**) 3D presentation of albumin, with highlighted fragments (*red*) exhibiting the greatest predisposition to adopt a beta form. (**C**) 200–400 aa fragment of Hb polypeptide chain. (**D**) Flexibility along the polypeptide chain in albumin as expressed by the RMS-F parameter

the dissociation constant and focuses electrostatic interactions on the polar regions. Consequently, rotation of the molecule about its long axis does not yield any benefits for self-association and can be considered irrelevant. Cis binding is likely related to the properties of other binding-capable substituents: the −OH and −NH$_2$ groups. In contrast, in CR the location of sulfonic groups clearly favors trans binding, i.e. alternating alignment of symmetrical halves (Fig. 3.10).

The structure of albumin suggests that the supramolecular ligand is attracted to the gap between the protein's two lobes [26]. This is where the longest non-helical folds can be found and where oscillatory structural changes (RMS-F) are revealed by molecular dynamic studies (depending on temperature), suggesting limited stability of the native secondary conformation. All these factors enable the anchoring of a supramolecular ligand (Fig. 3.11).

3.3 Fluorescence Property of Congo Red

Interesting conclusions may be drawn from fluorescence analysis. Free CR is generally not fluorescent; however, fluorescence appears when the dye forms complexes with cellulose or with certain proteins. Theoretical analysis indicates that fluorescence should be inhibited by self-association which increases the mobility of Π-electrons and besides does not produce dipoles of equal length. In turn, fluorescence should be expected to emerge when the micelle dissociates into individual molecules or (possibly) oligomers. The exact conditions which favor fluorescence are not known with certainty. Supramolecular CR resembles a twisted tape. In the absence of a complementary surface (which mirrors its twists), the ribbon may only adhere to other molecules locally. This disrupts the micelle and favors retention of individual molecules or oligomers, explaining the fluorescence observed when CR interacts with cellulose. Another evidence of the proposed mechanism is induced fluorescence which occurs when CR is dissociated by a detergent, such as cholate, which can be intercalated into the micelle due to its planar structure, but which lacks aromatic rings (Fig. 3.12).

Calculations [36] based on the assumption that the bond between CR and beta polysaccharides such as cellulose is mediated by the polar fragments of the dye, appear to suggest that the optimal arrangement would involve clustering of individual dye molecules perpendicular to the polysaccharide chain. Although this

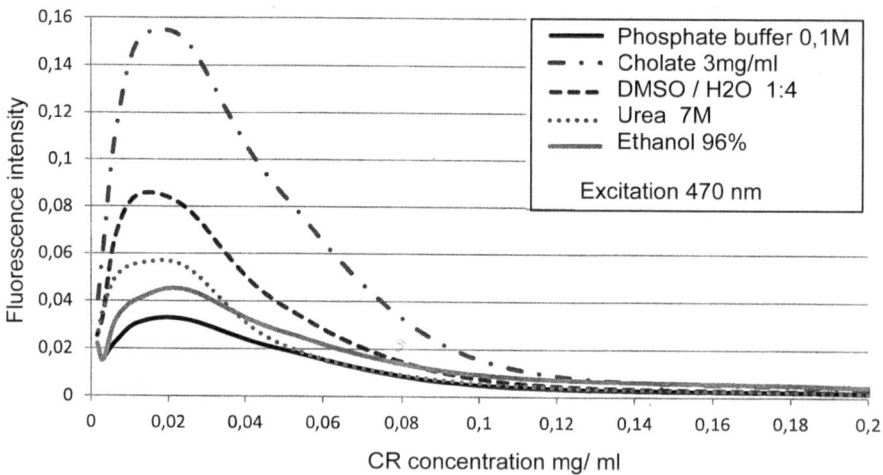

Fig. 3.12 Fluorescence of CR, arising as a result of disintegration of its micellar structure by DMSO (*dashed line*), alcohols (*solid gray line*) urea (*dotted line*) and cholate (*mixed dashed/dotted line*)

argument does not acknowledge the self-association tendencies of CR, both theories do account for the retention of individual CR molecules, which would explain the observed fluorescence. Similar dissociation of the micelle may also be observed when CR forms complexes with certain proteins or protein aggregates – e.g. immunoglobulins or amyloids. It may also be caused by mechanical strain and dislocations in proteins which form complexes with the supramolecular dye. Such a situation may arise e.g. when cells react via their surface receptors – mostly cadherins and integrins, which have immunoglobulin-like conformations and become susceptible to CR penetration as a result of structural strain [37]. Subsequent motion of the cells exacerbates tension and promotes structural rearrangements, which may fragment the attached supramolecular ligand [38]. An example is provided by the reaction between monocytes and cancer cells. Monocytes recognize cancer cells as alien and attack them. Figure 3.13 presents visible-light and UV images of this process. Fluorescent patches correspond to the specific areas where cancer cells are attacked by monocytes. The images reveal free monocytes with no signs of fluorescence, as well as aggregations of monocytes clustered around cancer cells, with CR fluorescence clearly visible. The presented interpretation is also supported by induction of fluorescence through structural strain and rearrangements in antibodies forming immune complexes. In order to reveal this effect, the rosetting technique has been applied [39].

This process involves monocytes which bind antibodies via their Fc receptors. Sheep red blood cells have been added to a solution of anti-SRBC antibodies. Subsequent images show monocyte rosettes entirely covered by erythrocytes, suggesting even distribution of the immune complex. This indicates that conditions which favor CR complexation may emerge anywhere on the cell surface. When analyzing UV images, it becomes evident that fluorescence appears where the structural strain is greatest, fragmenting the ligand micelle into smaller units and/or individual molecules (Fig. 3.14).

The fluorescence of CR induced in a rosette system indicates that the supramolecular ligand is destabilized and partly dissociated by structural strain in its attached protein which does not provide a close match for the ligand's own preferred conformation [40]. Similar results are obtained when heating complexes of CR with IgG light chains, which usually form two distinct fractions – the slow-moving and the fast-moving fraction. The former fraction is represented by ligands comprised of four dye molecules, whereas the latter fraction includes ligands with 5–8 dye molecules per micelle. There is only minimal smearing between the pure light chain stain and the slow-moving fraction stain, which indicates that slow-moving complexes are produced with ease. This fraction exhibits weak fluorescence. In contrast, the fast-moving fraction is formed slowly (preferentially at higher temperatures – 45–55 °C) and its electrophoretic image is smeared, indicating that the complex

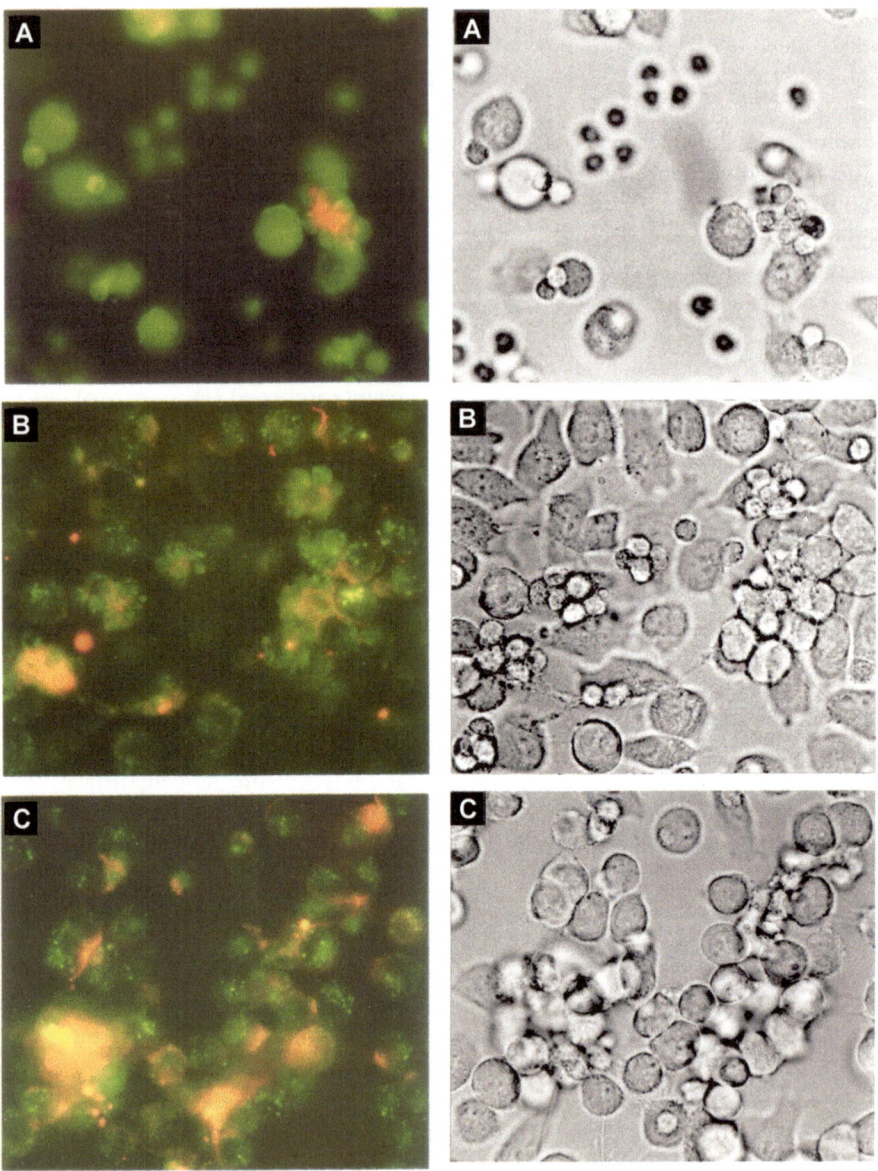

Fig. 3.13 CRfluorescence induced by cell-cell interaction (interacting monocytes and cancer cells) seen in UV (*left column*) and visible light (*right column*) – microscope imaging **A, B, C** - three independent examples. (Reproduced by permission of *Folia Histochemica et Cytobiologica*)

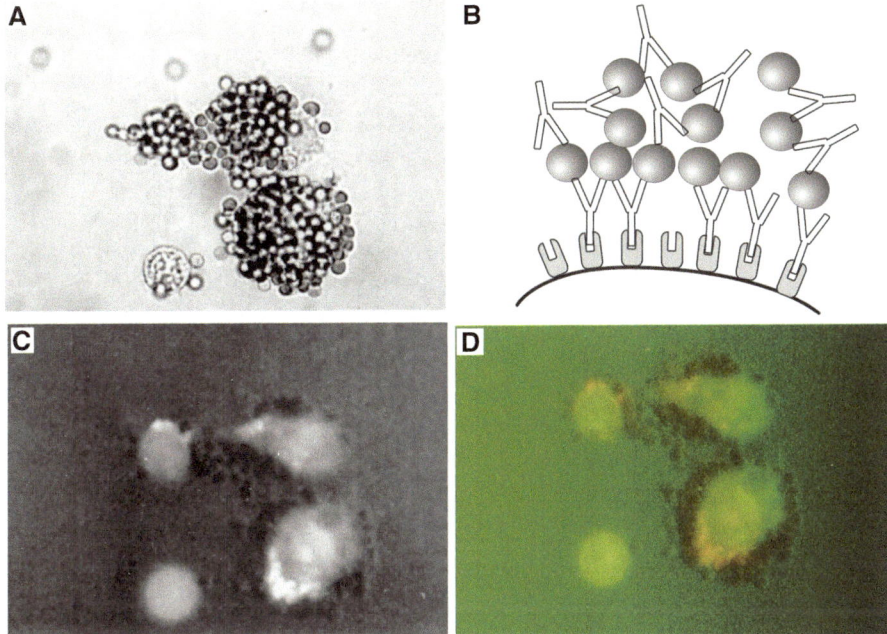

Fig. 3.14 Unequal CR fluorescence distribution in rosettes seen in visible light. (**A**) Visible light. (**B**) Simplified model of dye complexation by antibodies in the immune system. (**C** and **D**) UV spectra (Reproduced by permission of *Folia Histochemica et Cytobiologica*)

grows steadily from 5 to 8 dye molecules conquering some resistance – a process which also progressively increases its fluorescence (Fig. 3.15). The proposed mechanism is therefore validated. At higher dye concentrations, the distinction between both fractions is less dependent on temperature. This is likely due to simultaneous formation of both types of complexes in an environment characterized by abundance of CR.

As already discussed, supramolecular CR attaches itself to beta folds or random coils. This process is promoted by instabilities in the protein molecule, and, additionally, the complexation capabilities of the dye increase along with its concentration. By binding a supramolecular ligand, the protein adapts its tertiary conformation to the micelle; however most of the structural rearrangements which enable binding occur in the dye itself, and are facilitated by the noncovalent interactions between individual dye molecules.

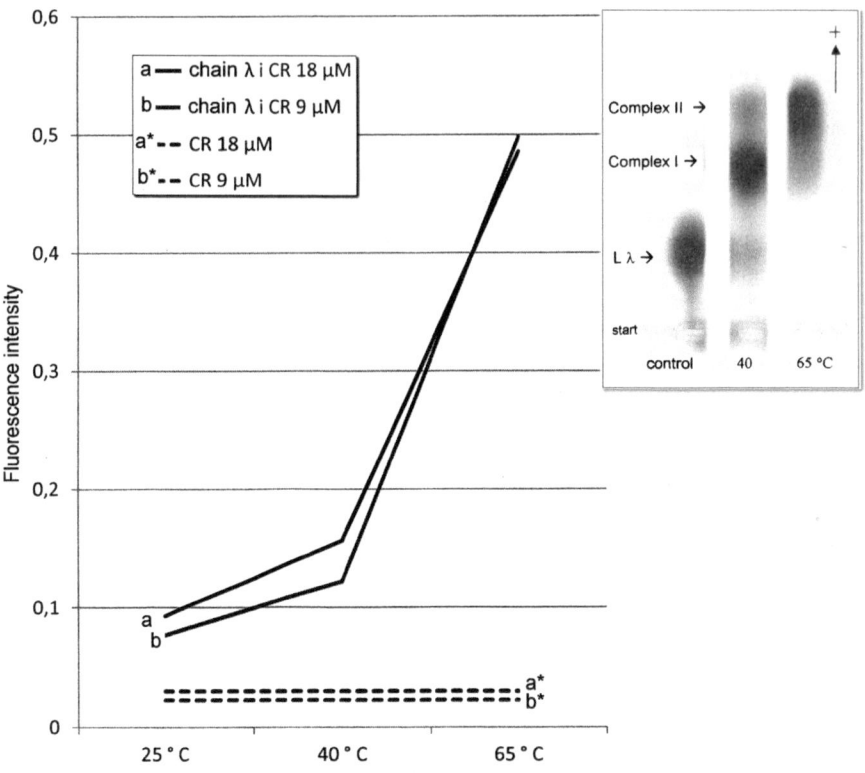

Fig. 3.15 Increase of fluorescence associated with formation of the L chain complexes with CR. *Inset*: temperature-dependent formation of corresponding complexes revealed under electrophoresis. *Dashed lines* – control CR

Acknowledgements We acknowledge the financial support from the National Science Centre, Poland (grant no. 2016/21/D/NZ1/02763) and from the project Interdisciplinary PhD Studies "Molecular sciences for medicine" (co-financed by the European Social Fund within the Human Capital Operational Programme) and Ministry of Science and Higher Education (grant no. K/DSC/001370).

References

1. Roterman I, No KT, Piekarska B, Kaszuba J, Pawlicki R, Rybarska J, Konieczny L (1993) Bis azo dyes –studies on the mechanism of complex formation with IgG modulated by heating or antigen binding. J Physiol Pharmacol 44(3):213–232
2. Rybarska J, Piekarska B, Stopa B, Zemanek G, Konieczny L, Nowak M, Król M, Roterman I, Szymczakiewicz-Multanowska A (2001) Evidence that supramolecular Congo red is the sole ligation form of this dye for L chain lambda derived amyloid proteins. Folia Histochem Cytobiol 39(4):307–314
3. Piekarska B, Konieczny L, Rybarska J, Stopa B, Zemanek G, Szneler E, Król M, Nowak M, Roterman I (2001) Heat-induced formation of a specific binding site for self-assembled Congo red in the V domain of immunoglobulin L chain lambda. Biopolymers 59(6):446–456

4. Lendel C, Bolognesi B, Wahlström A, Dobson CM, Gräslund A (2010) Detergent-like interaction of Congo red with the amyloid beta peptide. Biochemistry 49(7):1358–1360
5. Frid P, Anisimov SV, Popovic N (2007) Congo red and protein aggregation in neurodegenerative diseases. Brain Res Rev 53(1):135–160
6. Wang Y, Liu Y, Deng X, Cong Y, Jiang X (2016) Peptidic β-sheet binding with Congo Red allows both reduction of error variance and signal amplification for immunoassays. Biosens Bioelectron 86:211–218
7. Buell AK, Dobson CM, Knowles TP, Welland ME (2010) Interactions between amyloidophilic dyes and their relevance to studies of amyloid inhibitors. Biophys J 99(10):3492–3497
8. Howie AJ, Brewer DB (2009) Optical properties of amyloid stained by Congo red: history and mechanisms. Micron 40(3):285–301
9. Khurana R, Gillespie JR, Talapatra A, Minert LJ, Ionescu-Zanetti C, Millett I, Fink AL (2001) Partially folded intermediates as critical precursors of light chain amyloid fibrils and amorphous aggregates. Biochemistry 40(12):3525–3535
10. Ding Q, Cecarini V, Keller JN (2007) Interplay between protein synthesis and degradation in the CNS: physiological and pathological implications. Trends Neurosci 30(1):31–36
11. van Galen P, Kreso A, Mbong N, Kent DG, Fitzmaurice T, Chambers JE, Xie S, Laurenti E, Hermans K, Eppert K, Marciniak SJ, Goodall JC, Green AR, Wouters BG, Wienholds E, Dick JE (2014) The unfolded protein response governs integrity of the haematopoietic stem-cell pool during stress. Nature 510(7504):268–272
12. Lin JH, Li H, Yasumura D, Cohen HR, Zhang C, Panning B, Shokat KM, Lavail MM, Walter P (2007) IRE1 signaling affects cell fate during the unfolded protein response. Science 318(5852):944–949
13. Kawaguchi S, Ng DT (2011) Cell biology. Sensing ER stress. Science 333(6051):1830–1831
14. Edelman GM, Gally JA (1962) The nature of Bence-Jones proteins. Chemical similarities to polypetide chains of myeloma globulins and normal gamma-globulins. J Exp Med 116:207–227
15. Nakano T, Matsui M, Inoue I, Awata T, Katayama S, Murakoshi T (2011) Free immunoglobulin light chain: its biology and implications in diseases. Clin Chim Acta 412(11–12):843–849
16. Leitzgen K, Knittler MR, Haas IG (1997) Assembly of immunoglobulin light chains as a prerequisite for secretion. A model for oligomerization-dependent subunit folding. J Biol Chem 272(5):3117–3123
17. Kaplan B, Livneh A, Sela BA (2011) Immunoglobulin free light chain dimers in human diseases. Sci World J 11:726–735
18. Charafeddine KM, Jabbour MN, Kadi RH, Daher RT (2012) Extended use of serum free light chain as a biomarker in lymphoproliferative disorders: a comprehensive review. Am J Clin Pathol 137(6):890–897
19. Stevens FJ, Myatt EA, Chang CH, Westholm FA, Eulitz M, Weiss DT, Murphy C, Solomon A, Schiffer M (1995) A molecular model for self-assembly of amyloid fibrils: immunoglobulin light chains. Biochemistry 34(34):10697–10702
20. Raynes JG, Eagling S, McAdam KP (1991) Acute-phase protein synthesis in human hepatoma cells: differential regulation of serum amyloid A (SAA) and haptoglobin by inteleukin-1 and interlekin-6. Clin. Exp Immunol 83:488–491
21. Spólnik P, Piekarska B, Stopa B, Konieczny L, Zemanek G, Rybarska J, Król M, Nowak M, Roterman I (2003) The structural abnormality of myeloma immunoglobulins tested by Congo red binding. Med Sci Monit 9(4):BR145–BR153
22. Molloy TP, Wilson CW (1980) Protein-binding of ascorbic acid: 1. Binding of bovine serum albumin. Int J Vitam Nutr Res 50:380–386
23. Perrin JH, Nelson DA (1972) Induced optical activity following the binding of warfarin, indomethacin, 4-hydroxycoumarin and salicylic acid to human serum albumin. Life Sci I 11(6):277–283
24. Raghupathy E, Peterson NA, Estey SJ, Peters T Jr, Reed RG (1978) Serum albumin stimulation of synaptosomal proline uptake: partial identification on the active site. Biochem Biophys Res Commun 85(2):641–646
25. Ray A, Reynolds JA, Polet H, Steinhardt J (1966) Binding of large organic anions and neutral molecules by native bovine serum albumin. Biochemistry 5(8):2606–2616

26. Stopa B, Rybarska J, Drozd A, Konieczny L, Król M, Lisowski M, Piekarska B, Roterman I, Spólnik P, Zemanek G (2006) Albumin binds self-assembling dyes as specific polymolecular ligands. Int J Biol Macromol 40(1):1–8
27. Peters T Jr (1996) The albumin molecule: its structure and chemical properties. In: Peters T Jr (ed) All about albumin. Academic, San Diego/New York/Boston/London/Sydney/Tokyo/Toronto, pp 9–78
28. Dobryszycka W (1997) Biological functions of haptoglobin--new pieces to an old puzzle. Eur J Clin Chem Clin Biochem 35(9):647–654
29. Kurosky A, Barnett DR, Lee TH, Touchstone B, Hay RE, Arnott MS, Bowman BH, Fitch WM (1980) Covalent structure of human haptoglobin: a serine protease homolog. Proc Natl Acad Sci U S A 77(6):3388–3392
30. McCormick DJ, Atassi MZ (1990) Hemoglobin binding with haptoglobin: delineation of the haptoglobin binding site on the alpha-chain of human hemoglobin. J Protein Chem 9(6):735–742
31. Andersen CB, Torvund-Jensen M, Nielsen MJ, de Oliveira CL, Hersleth HP, Andersen NH, Pedersen JS, Andersen GR, Moestrup SK (2012) Structure of the haptoglobin-haemoglobin complex. Nature 489(7416):456–459
32. Ghozlane A, Joseph AP, Bornot A, de Brevern AG (2009) Analysis of protein chameleon sequence characteristics. Bioinformation 3(9):367–369
33. Spólnik P, Konieczny L, Piekarska B, Rybarska J, Stopa B, Zemanek G, Król M, Roterman I (2004) Instability of monoclonal myeloma protein may be identified as susceptibility to penetration and binding by newly synthesized Congo red derivatives. Biochimie 86(6):397–401
34. Konieczny L, Piekarska B, Rybarska J, Skowronek M, Stopa B, Tabor B, Dabroś W, Pawlicki R, Roterman I (1997) The use of Congo red as a lyotropic liquid crystal to carry stains in a model immunotargeting system--microscopic studies. Folia Histochem Cytobiol 35(4):203–210
35. Skowronek M, Stopa B, Konieczny L, Rybarska J, Piekarska B, Szneler E, Bakalarski G, Roterman I (1998) Self-assembly of Congo red – a theoretical approach to identify its supramolecular organization In water and salt solutions. Biopolymers 46:267–281
36. Woodcock S, Henrissat B, Sugiyama J (1995) Docking of Congo red to the surface of crystalline cellulose using molecular mechanics. Biopolymers 36(2):201–210
37. Burdick MM, McCarty OJ, Jadhav S, Konstantopoulos K (2001) Cell-cell interactions in inflammation and cancer metastasis. IEEE Eng Med Biol Mag 20(3):86–91
38. Zembala M, Siedlar M, Ruggiero I, Wieckiewicz J, Mytar B, Mattei M, Colizzi V (1994) The MHC class-II and CD44 molecules are involved in the induction of tumor necrosis factor (TNF) gene expression by human monocytes stimulated with tumour cells. Int J Cancer 56(2):269–274
39. Zupanska B (1990) Phenotypic markers for the feline monocyte: rosette which binds myeloid cells. Proc Soc Exp Biol Med 197:317–325
40. Roterman I, Rybarska J, Konieczny L, Skowronek M, Stopa B, Piekarska B, Bakalarski G (1998) Congo red bound to α-1-proteinase inhibitor as a model of supramolecular ligand and protein complex. Comput Chem 22:61–70

Chapter 4
Metal Ions Introduced to Proteins by Supramolecular Ligands

Olga Woźnicka, Joanna Rybarska, Anna Jagusiak, Leszek Konieczny, Barbara Stopa, and Irena Roterman

Abstract Congo red and other supramolecular structures may intercalate various foreign compounds, particularly planar ones. Such hybrid ligands, acting as a unit, may attach themselves to proteins and penetrate into their interior, together with any intercalated substances. If the intercalant is a metal complexone, a stable metallo-protein may be formed. This chapter discusses intercalation of metal complexones with metal ions bound by supramolecular Congo red as a means of introducing contrast to amyloid-like aggregates in order to trace the initial stages of amyloido-genesis. We investigate the applicability of Titan yellow carrying silver ions, and the alizarin complexone carrying tungsten and lead ions.

Keywords In vitro formed metalo-proteins • Supramolecular systems as metal ions carriages • Congo red and metal ions incorporation to proteins • Metal bearing supramolecular EM stains • Immune-complexes as Congo red target • Amyloid-like aggregates • EM contrast bearing Congo red • Metal ion complexation • Congo red and beta amyloid fibrils stained • Supramolecular systems and metal ion complexation

O. Woźnicka (✉)
Department of Cell Biology and Imaging, Jagiellonian University,
Gronostajowa 9, 30-387 Krakow, Poland
e-mail: olga.woznicka@uj.edu.pl

J. Rybarska • A. Jagusiak • L. Konieczny • B. Stopa
Chair of Medical Biochemistry, Jagiellonian University – Medical College,
Kopernika 7, 31-024, Krakow, Poland
e-mail: mbstylin@cyf-kr.edu.pl; anna.jagusiak@uj.edu.pl; mbkoniec@cyf-kr.edu.pl;
barbara.stopa@uj.edu.pl

I. Roterman
Department of Bioinformatics and Telemedicine, Jagiellonian University – Medical College,
Łazarza 16, 31-530, Krakow, Poland
e-mail: myroterm@cyf-kr.edu.pl

© The Author(s) 2018
I. Roterman, L. Konieczny (eds.), *Self-Assembled Molecules – New Kind of Protein Ligands*, https://doi.org/10.1007/978-3-319-65639-7_4

4.1   Metal Ions in Natural Biological Systems

Many proteins, including enzymes, rely on metal ions for their biological activity. The ions themselves are usually transition metals, such as iron, cobalt, nickel or copper. They enable catalysis due to their electron structure and the ability to form coordinate bonds. Another commonly encountered metal is zinc – it can be found e.g. in proteolytic enzymes known as metalloproteinases whose peculiar complexation capabilities have attracted much scientific attention.

Nearly one-third of all known enzymes include some type of metal ion. Metals are primarily associated with catalysis, but they also play an important part in formation of specific complexes, such as oxygen binding – e.g. in hemoglobin, which contains iron. Metals are also encountered in transcription factors, and along many other biochemical pathways, such as respiration. Certain protein complexes are dedicated to sequestration and/or accumulation of metals (e.g. siderophilin and ferritin) as well as detoxification (e.g. metallothionein).

Transition metals are sometimes referred to as "d"-electron metals due to the involvement of their "d" orbitals in atomic interactions. They form a variety of compounds with interesting spectroscopic and magnetic properties whose practical applications are the subject of ongoing research [1–17]. In the cell, metals are usually found in their complexed form, either as standalone ions or inside specific planar carrier compounds encapsulated by proteins – e.g. the porphyrin ring in hemoglobin. Proteins provide the capability to bind metals, isolate them from water and may introduce favorable steric conditions [18]. A classic example is hemoglobin, which enables oxygen binding without oxidation of the bivalent iron. Combining metal ions with proteins enhances their catalytic potential. The search for artificial structures with desirable catalytic properties is an important topic in medical science. The bond between the metal and the protein should not be random in character, since such uncontrolled complexation is usually encountered on the protein surface, where the metal remains in contact with water. In order to reduce polarity, the ion should optimally be placed right in the pocket of the active group. In practice, however, this is a very challenging task. An interesting solution is proposed in [19], where a metal ion was attached to streptavidin by connecting the Ru complex with biotin via a carboxyl group. This resulted in catalytic activity even though the complexed ion was located nearly on the surface (the biotin binding cavity has a depth of approximately 15Å, which coincides with the length of the complex) [20].

4.2   Insertion of Metal Ions into Proteins by Supramolecular Ligands

An entirely different approach to binding metals with proteins relies on supramolecular ligands. Such ligands can form stable complexes with proteins which have become susceptible to penetration as a result of function-related structural changes, or which possess such properties natively [21–23]. Certain proteins can be induced to undergo complexation with supramolecular ligands e.g. by heating, which causes partial unfolding of the polypeptide chain. However, only ribbonlike supramolecular structures are capable of penetrating into proteins and forming stable complexes. Examples include CR, EB and others [24, 25] – dyes with known affinity for amyloids and immune complexes [26]. Supramolecular systems (e.g. CR) may, in turn, intercalate a variety of planar compounds with a polyaromatic ring structure and/or positive charge. Hence if the intercalant itself contains a metal ion, the ion can be attached to the protein. As a result supramolecular ligands can bind metal ions to proteins by intercalating their complexons.

The supramolecular ligand usually binds outside of the protein's active site. The complex is formed in the distal part of the molecule, however the ligand penetrates into the protein interior, where polarity is lower and no water is present – similarly to the active site [22, 27, 28]. Exposure of the intercalated metal ion depends on the structure of the resulting complex. This means that attaching metals to proteins seems at the moment more convenient than through the use of customized substrates or enzymatic inhibitors. The ion is not delivered to the active site, but the resulting conformation shares some similarities with the structure of the active site, and may possess useful biological properties.

The goal of our team was to confirm the proposed means of attaching metals to proteins, and also to devise a way to equip CR – a known amyloid stain – with contrast for the purposes of EM imaging. The complexation of CR with amyloids is the subject of numerous studies [29–31]. The problem is difficult and its molecular underpinnings remain speculative, since CR itself is not visualized under electron microscopy, while amyloids – despite their ordered structure – do not attain the necessary level of crystallization order. One putative solution would be to add contrast to CR itself to visualize its distribution with respect to amyloid deposits. Assuming that CR binds to amyloids as a supramolecular system, the contrast could be introduced as a metal-containing intercalant. The proposed compound – TY – comprises symmetric polar groups and aromatic rings [32, 33] (Fig. 4.1). Its halves are linked by a tri-azo bond capable of complexing metal ions, particularly silver and mercury. The silver-containing complex is more convenient due to its stability and formation in both neutral and slightly alkaline environments. TY/Ag$^+$ complexation is also easy to detect since its spectrum differs markedly from the spectrum of free TY (Fig. 4.1).

The complex withstands electrophoretic dissociation in alkaline pH in a tris buffer (Fig. 4.2). It dissociates in the presence of anions, yielding insoluble silver compounds (solubility coefficient $< 10^{-13}$), which indicates that in biological systems only thiol groups are effectively able to react with silver bound in the complex. A certain disadvantage of this complex is its notable viscosity. This is due to the specific mode of interaction between the silver ion and the tri-azene bond – it may involve both coordination valence electrons of silver ion, or only one, if the other has easy access to the tri-azene group of another dye molecule. This phenomenon promotes association of complexes, particularly at high dye concentrations. Fortunately, the viscosity of TY/Ag$^+$ complexes is greatly decreased in the presence

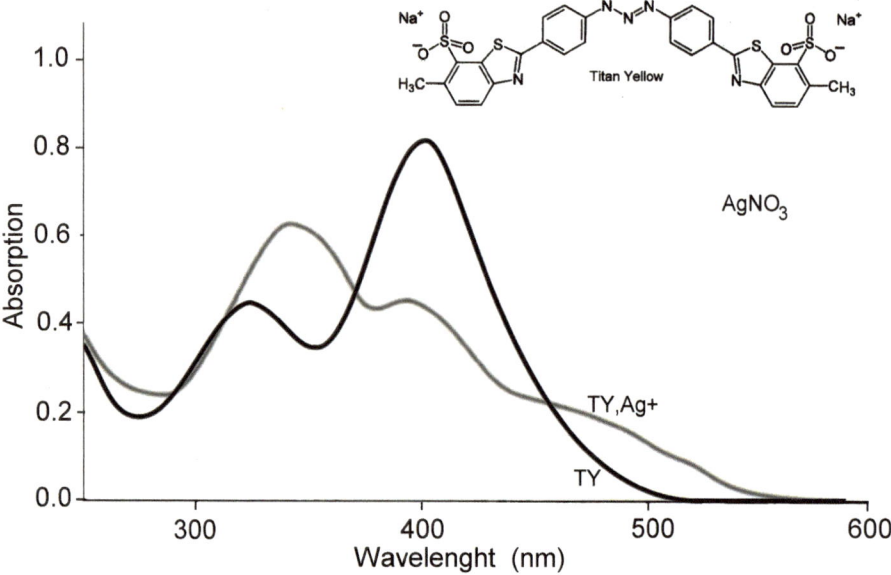

Fig. 4.1 Spectral changes effect of silver ion complexation with TY
Inset: the formula of TY

Fig. 4.2 Agarose
electrophoresis of dyes
used for staining amyloids
and amyloid-like
aggregates. Accelerated
migration of CR mixed
with TY is the evidence of
mutual complexation.
1 – TY/Ag⁺, *2* – TY,
3 – CR/TY, *4* – CR

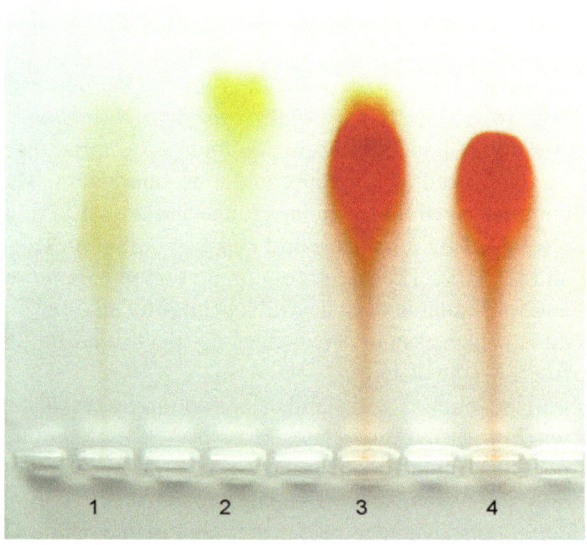

of CR, whose molecules separate the intercalants, preventing mutual interactions. Intercalation of TY enhances the contrast characteristics of CR, rendering it useful for EM imaging of amyloids.

Initial research involving contrast-enhanced CR has been carried out with amyloid-like aggregations formed by shaking of IgG light chains at increased temperatures (40–45 °C), following which the precipitate was treated with the CR-TY-Ag$^+$ complex. The progressive appearance of structural order can be observed by imaging CR, which is a selective amyloid stain. In the presented case, aggregations were formed by shaking an amyloidogenic protein (IgG light chain) in conditions verging on denaturation: increased temperature and pH inconsistent with the protein's isoelectric point. These conditions promote the formation of amorphous aggregations in which seeds of order randomly emerge – typically as short, spiraling threads capable of binding the supramolecular dye. Such ordering is easier to observe and analyze along the edges of solids and/or in fine flecks of aggregates. The contrast medium carried by CR greatly simplifies molecular analysis.

Figure 4.3 presents the contrast-enhanced aggregate as seen under an electron microscope. Accurate interpretation of results requires however carefully prepared samples and an optimized method – still, even at this early stage it seems evident that the image carries useful information. Functional optimization should involve formation and stability of the protein-metal complex, but also even distribution of CR and the intercalated contrast medium. A classic amyloid aggregation formed by commercially available peptides (amyloid beta) is shown in Fig. 4.4 – it appears as twisted strands whose thickness depends on the synthesis procedure.

The proposed approach may also be used to stain antibodies engaged in immune complexes, since immune complexation renders antibodies susceptible to penetration by CR [34, 35]. Figure 4.5 presents sheep red blood cell membranes (ghosts) broken to pieces by homogenization and then agglutinated by specific antibodies obtained from the sensitized rabbit (anti-SRBC serum).

Another, more universal complexon analyzed in the search for convenient methods of attaching metal ions into proteins is the alizarincomplexone (Fig. 4.6). It fulfills all the necessary conditions – it can bind a variety of metals and is itself intercalated by CR.

Unfortunately, it requires acidic conditions, since metal complexation is mediated by two acetic acid residues (which undergo ionization in a neutral or basic environment, declining their ability to bind the metal). The problem is additionally exacerbated by the fact that in pH < 5 CR transitions into a chinoid compound, becoming insoluble. Taken together, these two phenomena establish a rather narrow pH range where effective complexation may take place. The alizarincomplexone forms complexes with cobalt, lead, nickel and with the salts of certain other metals such as molybdenum and nickel. Complexation is evidenced by spectral changes (Figs. 4.7 and 4.8). Complexes persist under electrophoresis.

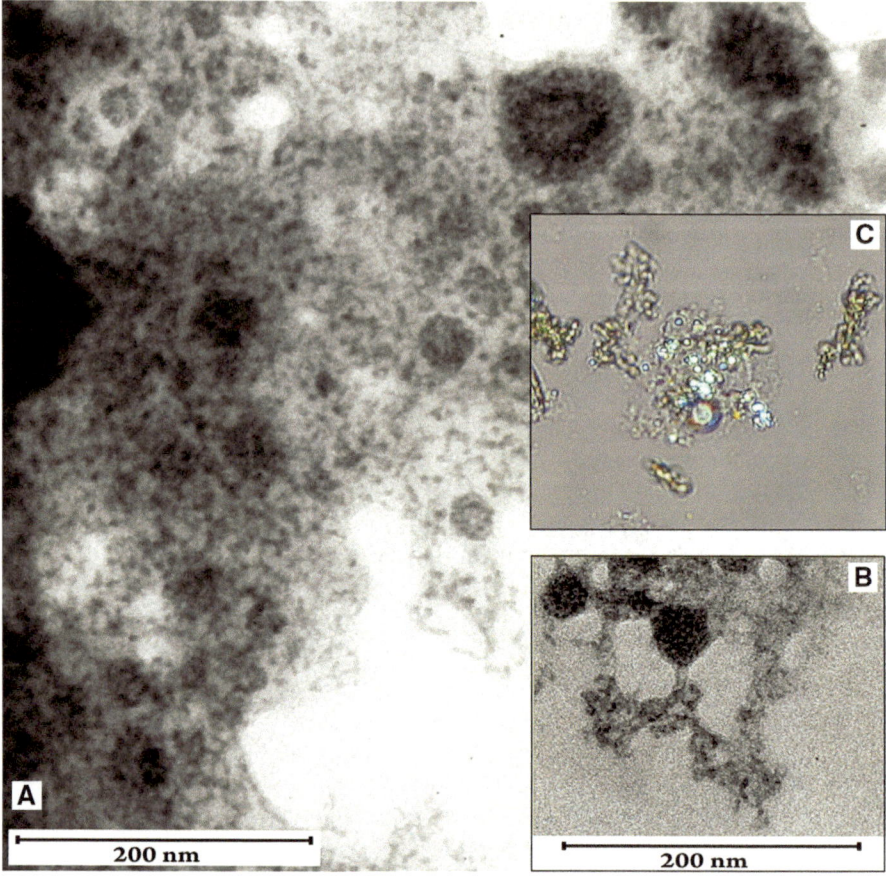

Fig. 4.3 Amyloid-like particles. (**A** and **B**) The edge fragments contrasted with complex CR/TY/Ag⁺ used as the stain. (**C**) Aggregates seen in the polarised light

Tungsten and lead complexes may serve as a contrast medium when intercalated by CR in 0.05 M acetate buffer (pH = 5.9).

The resulting EM images reveal various modes of aggregation, depending on the target polypeptide chain and environmental conditions. In most cases the aggregate is amorphous but contains islands of ordered structures, seen as granular chains of varying length, with longer chains often spirally twisted. When viewed in polarized light, these ordered fragments exhibit birefringence and glow, indicating that they constitute of ordered amyloid precursors (Figs. 4.9, 4.10, 4.11 and 4.12). Nevertheless, a detailed description of their structure requires further research.

Fig. 4.4 Amyloid Beta
1–40 (Sigma) derived
fibrils stained by (CR/TY/
Ag⁺) complex seen in EM
picture. Supplement
picture below – the
enlargement of fibrils

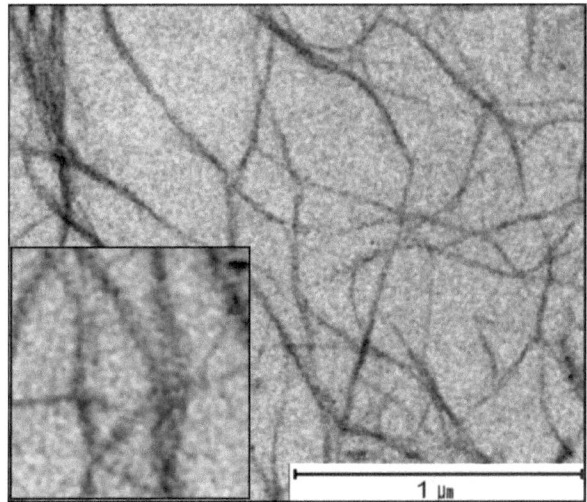

4.3   Conclusions

Transition metal ions exhibit complexation capabilities which may be of significant
practical importance. Biological applications focus on introduction of such ions into
proteins, where they exhibit greater reactivity than in their unbound form. The role
of the protein in this process is not entirely clear, although changes in environmental
factors (particularly reduction in polarity) appear to play an important role. Under
these assumptions, it is not necessary for the metal ion to be placed in the active site,
since this is usually not the only area characterized by low polarity. Internalization
of metal ions is facilitated by supramolecular ligands, capable of penetrating into
proteins in unstable areas basically other than the active site. Such ligands can serve
as carriers for a variety of organic molecules, including those which contain metal
ions. The metal-containing compound is attached to the ligand via intercalation,
which naturally leads us to search for suitable metal complexones, able to form
strong bonds with CR. Examples include TY and the alizarin complexone.
Confirmation of the metal-protein bond is provided by microscopic imaging of
amyloid-like aggregations formed by the IgG light chain stained by CR in complex
with TY/silver ions.

EM images reveal amorphous aggregations which contain ordered structures
capable of binding the contrast medium. Such structures typically adopt the form of

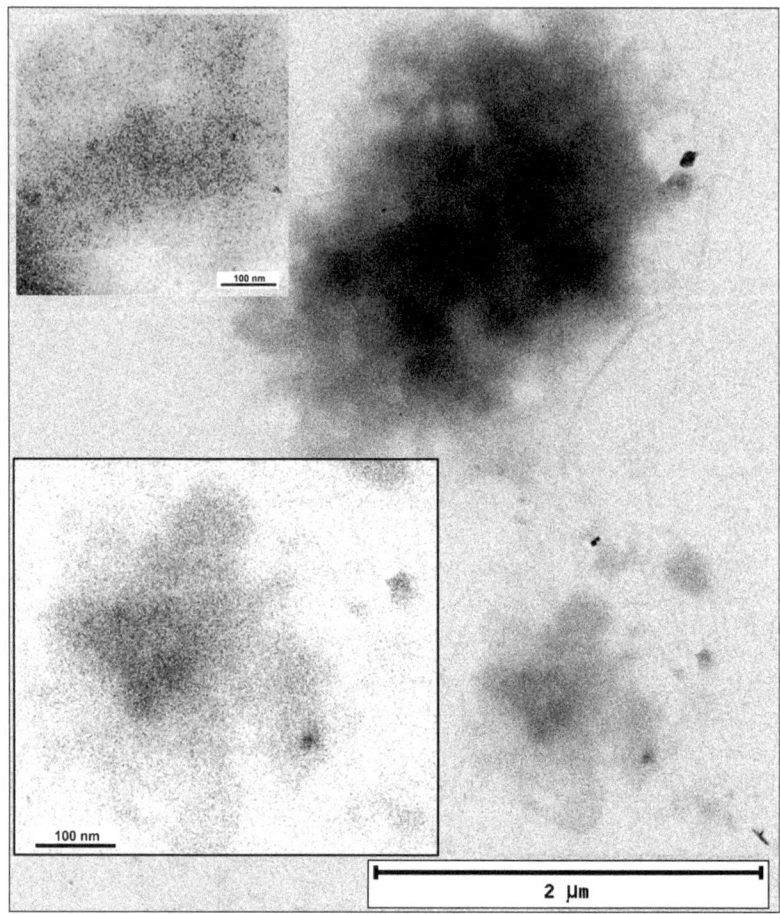

Fig. 4.5 Agglutinates of fragmented by homogenisation erythrocyte ghosts (SRBC) stained by CR/TY/Ag⁺. *Insets* – enlarged fragments

twisted strands and function as seeds for amyloid transformation. In addition to bulk solids (which are difficult to study with EM), some aggregates also adopt the form of thin membranes, exposing the contrast medium and confirming that supramolecular ligands may successfully attach metal ions to proteins. Thus, the location of CR concentrations in microscopic images reveals the shape of the analyzed object. While amyloids are known to adopt various structures, depending on the conformation of the unit protein and on environmental conditions, all such structures retain the ability to bind CR (Figs. 4.7, 4.8, 4.9 and 4.10) show fragments of amyloid-like aggregates with portions containing ordered structures exposed by contrast. As can be seen, the bulk of the aggregate remains amorphous and incapable of

A

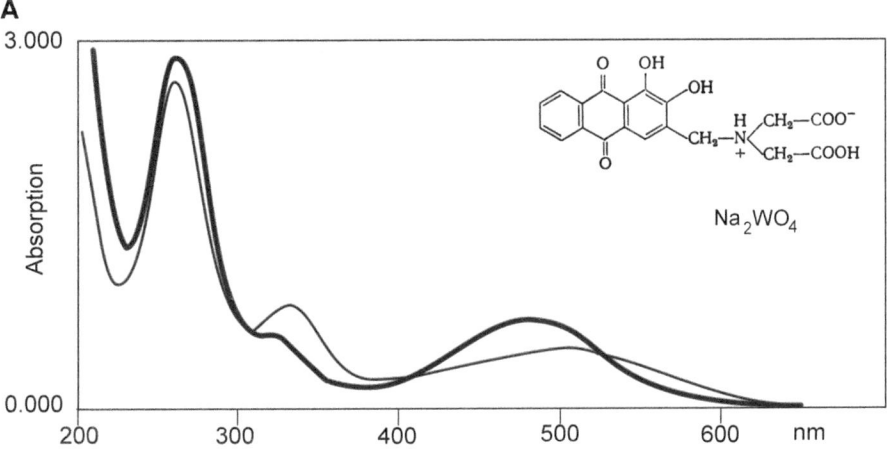

B

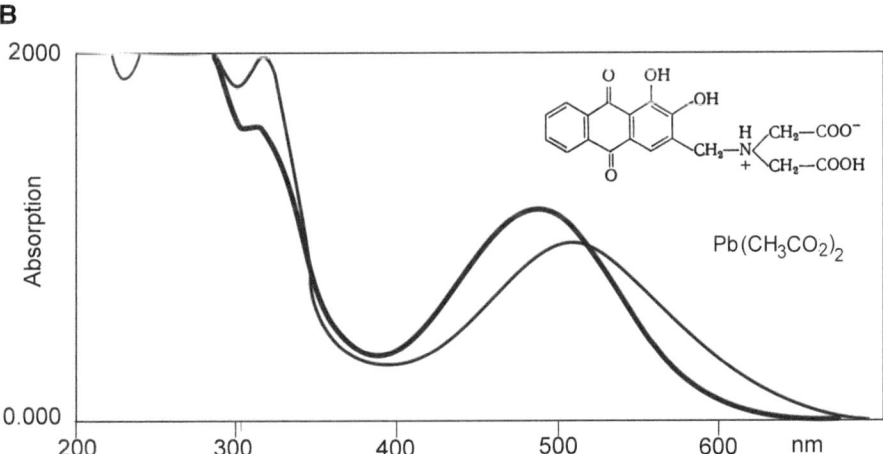

Fig. 4.6 Spectral changes associating the complexation of alisarine complexon (AC) with tungstate ion – **A** and lead ion – **B**

binding the dye. Due to its thickness, it produces some darkening in the resulting images, but even so the contrast is clearly discernible. The ordered forms constitute seeds for amyloid transformation, which explains their ability to bind CR.

Aggregation is mediated by light chain V domains, which are less stable than the corresponding C domains. Prior to aggregation, V domains undergo partial unfolding initiated by displacement of N-terminal chain. This uncovers the interior of domains [21] making available a pocket in which the CR aggregate may anchor itself. It appears that unfolding affects the so-called upper core [36, 37] and occurs symmetrically in the dimeric structure of the protein. Partly unfolded V domains may subsequently form beta-beta bonds with adjacent molecules, creating a fibril.

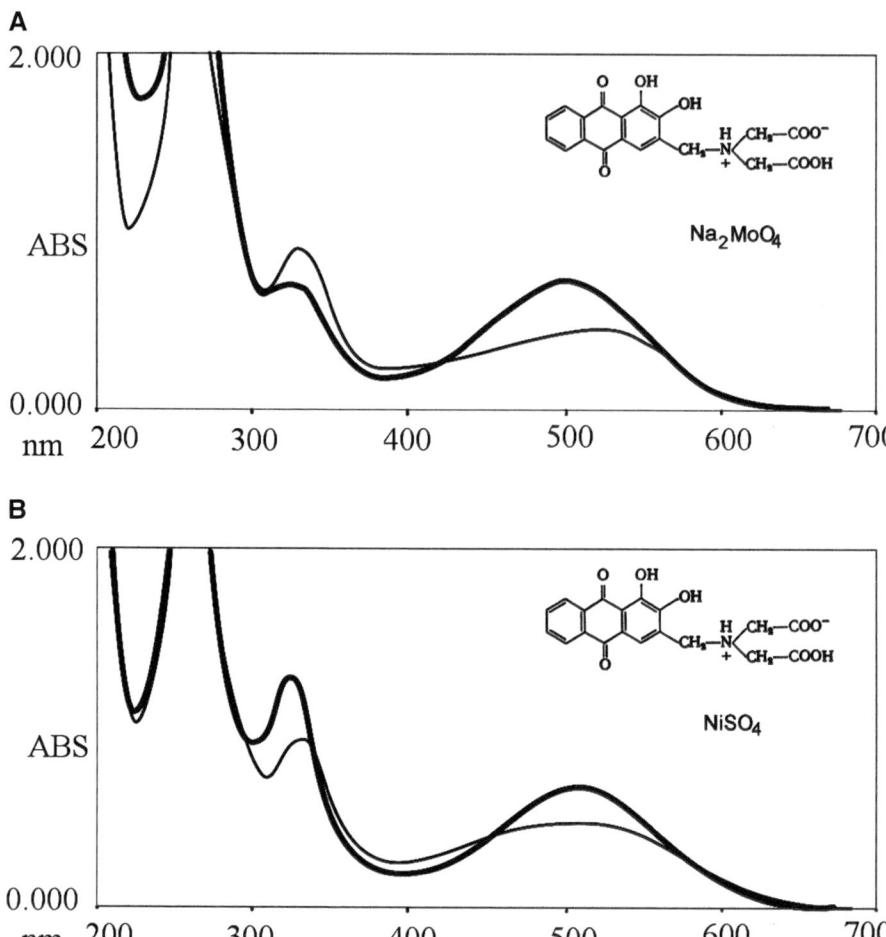

Fig. 4.7 Spectral changes presented as an evidence of alizarin complexon binding different metals – Mo (**A**) as molibdate MoO_4^{2-} and Ni^{2+} (**B**)

The role of the lower core (responsible for stabilization of the inter-chain interface) in this process is not clear, as is its involvement in structural rearrangements in the V domain. It seems that fibrils can form in either case; however, strand-like products are expected to possess different properties. Molecular interpretation of CR binding to amyloid fibrils formed by commercially available Amyloid Beta peptide 40 requires further study. In contrast to native proteins which undergo structural rearrangements facilitating penetration of a supramolecular ligand, the packing of Beta peptide 40 derived amyloid fibrils is tight and uniform. The dye may potentially penetrate in areas where the fibril is sheared or otherwise structurally disrupted.

A

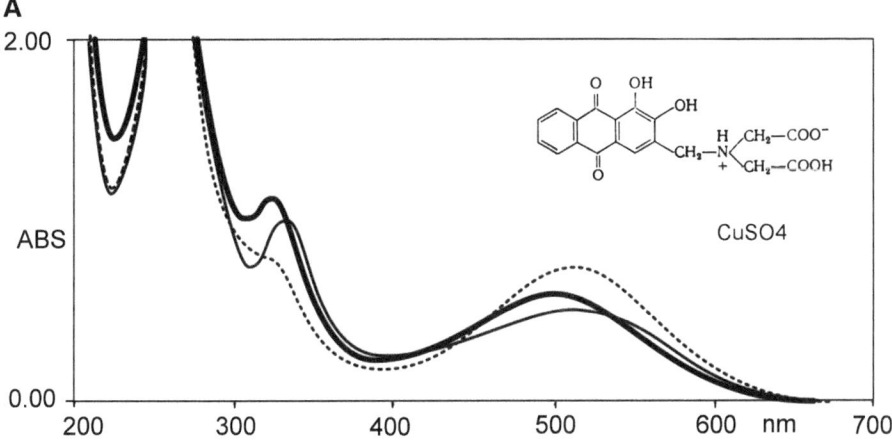

B

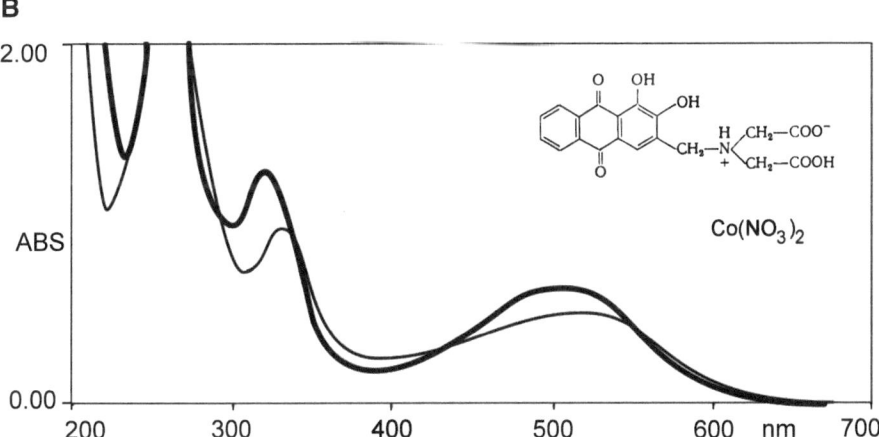

Fig. 4.8 Spectral changes presented as an evidence of alizarincomplexon binding ions of different metals – Cu (**A**) and Co (**B**). The spectrum change marked by *dotted line* indicates that Cu^{2+} complexation reveals the formation of modified complex at the excess of Cu^{2+} ions added. Molar ratio of AC: Cu^{2+} 1:2

Complexation is further hampered by the fact that fibrils often consist of multiple intertwined strands – in such cases the ligand may engage individual strands without disrupting the fibril as a whole. This theory is supported by EM analysis, showing that CR does not uniformly cover the entire fibril, despite an abundance of the dye (Fig. 4.4), and that the fibril is unwound in many areas. It should also be noted that ribbon-like CR may attach itself to amyloids by means of adhesion.

Immune complexes are another type of object which can be visualized in the same manner – this is due to the fact that, by binding their natural ligand, antibodies

Fig. 4.9 Islands of ordered
structures formed within
the amyloid-like particle
stained selectively with
CR/AC/WoO$_4^{2-}$ complex

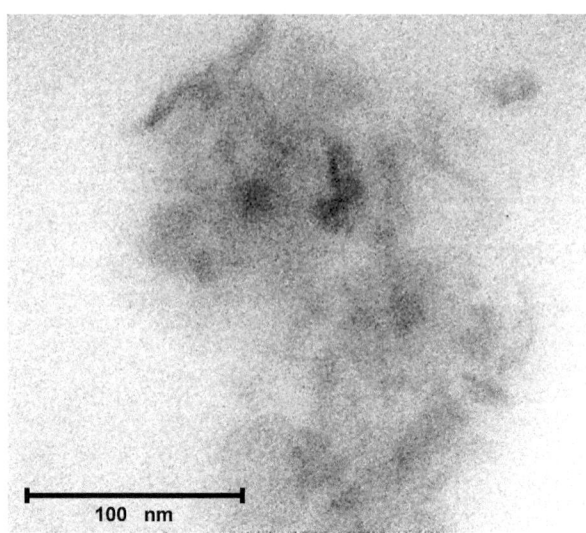

Fig. 4.10 Amyloid-like
particles with ordered
structures inside stained
with CR/AC/Pb^{+2}

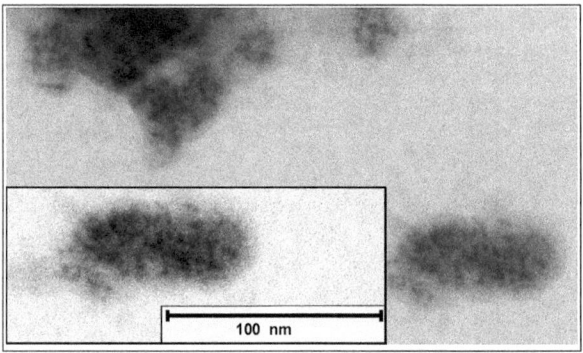

incur local structural instabilities which render them susceptible to penetration by supramolecular dyes, such as CR.

The proposed ligands and the procedure outlined in this work are both preliminary in scope. The technique can be further optimized by carefully selecting reagents and modifying the manner in which metal ions are transported by supramolecular ligands.

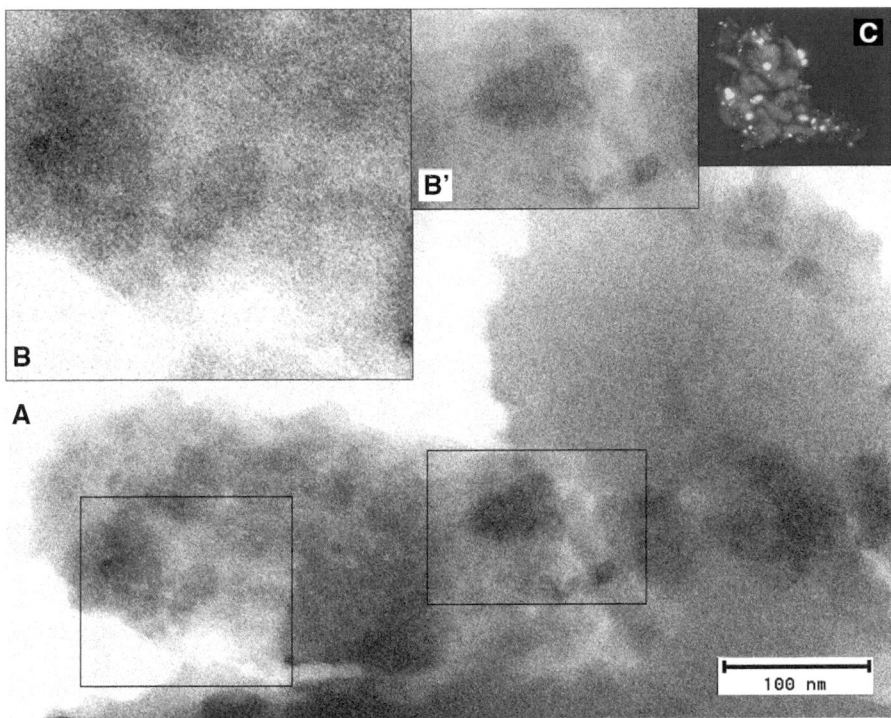

Fig. 4.11 Islands of ordered structure within amyloid-like particle stained selectively with (**A**) CR/AC/Pb^{2+} complex used as the contrast for EM studies. (**B** and **B'**) - enlarged fragments, (**C**) - islands of ordered structures seen in amyloid-like particle in polarised light

Fig. 4.12 The ordered structures bearing fragment outgrowing unstructured amyloid particle stained by CR/AC/Pb^{+2}. Local burst of ordered structures

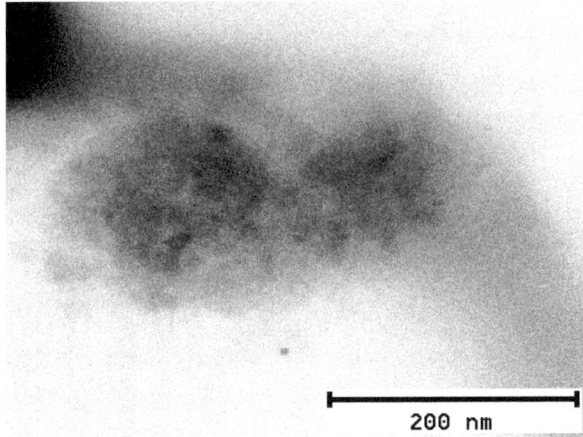

Acknowledgements We acknowledge the financial support from the National Science Centre, Poland (grant no. 2016/21/D/NZ1/02763) and from the project Interdisciplinary PhD Studies "Molecular sciences for medicine" (co-financed by the European Social Fund within the Human Capital Operational Programme) and Ministry of Science and Higher Education (grant no. K/DSC/001370).

References

1. Sorenson JR, Wangila GW (2007) Co-treatment with copper compounds dramatically decreases toxicities observed with cisplatin cancer therapy and the anticancer efficacy of some copper chelates supports the conclusion that copper chelate therapy may be markedly more effective and less toxic than cisplatin therapy. Curr Med Chem 14(14):1499–1503
2. Begley TP (2017) Origin of a key player in methane biosynthesis. Nature 543(7643):49–50
3. Biersack B, Ahmad A, Sarkar FH, Schobert R (2012) Coinage metal complexes against breast cancer. Curr Med Chem 19(23):3949–3956
4. Cooper GJ (2012) Selective divalent copper chelation for the treatment of diabetes mellitus. Curr Med Chem 19(17):2828–2860
5. Zhang J, Zhang F, Li H, Liu C, Xia J, Ma L, Chu W, Zhang Z, Chen C, Li S, Wang S (2012) Recent progress and future potential for metal complexes as anticancer drugs targeting G-quadruplex DNA. Curr Med Chem 19(18):2957–2975
6. Wang T, Guo Z (2006) Copper in medicine: homeostasis, chelation therapy and antitumor drug design. Curr Med Chem 13(5):525–537
7. Ashfaq M, Najam T, Shah SS, Ahmad MM, Shaheen S, Tabassum R, Rivera G (2014) DNA binding mode of transition metal complexes, a relationship to tumor cell toxicity. Curr Med Chem 21(26):3081–3094
8. El Kazzouli S, El Brahmi N, Mignani S, Bousmina M, Zablocka M, Majoral JP (2012) From metallodrugs to metallodendrimers for nanotherapy in oncology: a concise overview. Curr Med Chem 19(29):4995–5010
9. Fanni D, Fanos V, Gerosa C, Piras M, Dessi A, Atzei A, Van EP, Gibo Y, Faa G (2014) Effects of iron and copper overload on the human liver: an ultrastructural study. Curr Med Chem 21(33):3768–3774
10. Gumienna-Kontecka E, Pyrkosz-Bulska M, Szebesczyk A, Ostrowska M (2014) Iron chelating strategies in systemic metal overload, neurodegeneration and cancer. Curr Med Chem 21(33):3741–3767
11. Inoue K, O'Bryant Z, Xiong ZG (2015) Zinc-permeable ion channels: effects on intracellular zinc dynamics and potential physiological/pathophysiological significance. Curr Med Chem 22(10):1248–1257
12. Kaluderović GN, Paschke R (2011) Anticancer metallotherapeutics in preclinical development. Curr Med Chem 18(31):4738–4752
13. Kellett A, Prisecaru A, Slator C, Molphy Z, McCann M (2013) Metal-based antimicrobial protease inhibitors. Curr Med Chem 20(25):3134–3151
14. Ruiz-Azuara L, Bravo-Gómez ME (2010) Copper compounds in cancer chemotherapy. Curr Med Chem 17(31):3606–3615
15. Tardito S, Marchiò L (2009) Copper compounds in anticancer strategies. Curr Med Chem 16(11):1325–1348

16. Terreno E, Dastrù W, Delli Castelli D, Gianolio E, Geninatti Crich S, Longo D, Aime S (2010) Advances in metal-based probes for MR molecular imaging applications. Curr Med Chem 17(31):3684–3700
17. Ward RJ, Dexter DT, Crichton RR (2012) Chelating agents for neurodegenerative diseases. Curr Med Chem 19(17):2760–2772
18. Collman JP, Brauman JI, Rose E, Suslick KS (1978) Cooperativity in O2 binding to iron porphyrins. Proc Natl Acad Sci U S A 75(3):1052–1055
19. Jeschek M, Reuter R, Heinisch T, Trindler C, Klehr J, Panke S, Ward TR (2016) Directed evolution of artificial metalloenzymes for in vivo metathesis. Nature 537(7622):661–665
20. Green NM, Konieczny L, Toms EJ, Valentine RC (1971) The use of bifunctional biotinyl compounds to determine the arrangement of subunits in avidin. Biochem J 125(3):781–791
21. Piekarska B, Konieczny L, Rybarska J, Stopa B, Zemanek G, Szneler E, Król M, Nowak M, Roterman I (2001) Heat-induced formation of a specific binding site for self-assembled Congo Red in the V domain of immunoglobulin L chain lambda. Biopolymers 59(6):446–456
22. Roterman I, No KT, Piekarska B, Kaszuba J, Pawlicki R, Rybarska J, Konieczny L (1993) Bis azo dyes--studies on the mechanism of complex formation with IgG modulated by heating or antigen binding. J Physiol Pharmacol 44(3):213–232
23. Stopa B, Piekarska B, Konieczny L, Rybarska J, Spólnik P, Zemanek G, Roterman I, Król M (2003) The structure and protein binding of amyloid-specific dye reagents. Acta Biochim Pol 50(4):1213–1227
24. Król M, Roterman I, Piekarska B, Konieczny L, Rybarska J, Stopa B, Spólnik P, Szneler E (2005) An approach to understand the complexation of supramolecular dye Congo red with immunoglobulin L chain lambda. Biopolymers 77(3):155–162
25. Skowronek M, Stopa B, Konieczny L, Rybarska J, Piekarska B, Szneler E, Bakalarski G, Roterman I (1998) Self-assembly of Congo red – a theoretical and experimental approach to identify its supramolecular orhanization in water and salt solutions. Biopolymers 46:267–281
26. Rybarska J, Konieczny L, Roterman I, Piekarska B (1991) The effect of azo dyes on the formation of immune complexes. Arch Immunol Ther Exp (Warsz) 39(3):317–327
27. Piekarska B, Konieczny L, Rybarska J (1988) The effect of azo dyes on heat aggregation of IgG. Acta Biochim Pol 35(4):297–305
28. Konieczny L, Piekarska B, Rybarska J, Stopa B, Krzykwa B, Nowowrolski J, Pawlicki R, Roterman I (1994) Bis-azo dye liquid crystalline micelles as possibile drug carriers in immunotargeting technique. J Physiol Pharmacol 45:441–454
29. Frid P, Anisimov SV, Popovic N (2007) Congo red and protein aggregation in neurodegenerative disease. Brain Res Rev 53:135–160
30. Lendel C, Bolognesi B, Wahlström A, Dobson CM, Gräslund A (2010) Detergent-like interaction of Congo red with the amyloid β-peptide. Biochemistry 49:1358–1360
31. Khurana R, Gillespie JR, Talapatra A, Minert LJ, Ionescu-Zanetti C, Millett I, Fink AL (2001) Partially folded intermediates as critical precursors of light chain amyloid fibrils and amorphous aggregates. Biochemistry 40:3525–3535
32. Lazny R, Poplawski J, Köbberling J, Anders D, Brësse S (1999) Triazenes: a useful protecting strategy for sensitive secondary amines. Synlett 8:1301–1306
33. King HGC, Pruden G (1967) The component of commercial titan yellow most reactive towards magnesium: Its isolation and use in determining magnesium in silicate minerals. Analyst 92:83–90
34. Rybarska J, Konieczny L, Roterman I, Piekarska B (1991) The effect of Azo dyes on the formation of immune complexes. Archiv Immunol Ther Exp 39:317–327

35. Jagusiak A, Konieczny L, Krol M, Marszalek P, Piekarska B, Piwowar P, Roterman I, Rybarska J, Stopa B, Zemanek G (2015) Intramolecular immunological signal hypothesis revived – structural background of signalling revealed by using Congo red as a specific tool. Mini Rev Med Chem 14(13):1104–1113
36. Ewert S, Honegger A, Plückthun A (2004) Stability improvement of antibodies for extracellular and intracellular applications: CDR grafting to stable frameworks and structure-based framework engineering. Methods 34(2):184–199
37. Röthlisberger D, Honegger A, Plückthun A (2005) Domain interactions in the Fab fragment: a comparative evaluation of the single-chain Fv and Fab format engineered with variable domains of different stability. J Mol Biol 347(4):773–789

Chapter 5
Possible Mechanism of Amyloidogenesis of V Domains

Mateusz Banach, Barbara Kalinowska, Leszek Konieczny, and Irena Roterman

Abstract This chapter discusses complexation of Congo red by amyloid structures comprising immunoglobulin light chains, particularly the so-called Bence-Jones (BJ) proteins. According to the presented study, in BJ proteins the V domain is substantially less stable than the C domain. This conclusion is based on quantitative analysis of the protein's hydrophobic core, made possible by the fuzzy oil drop model. Results indicate that the V domain exhibits structural ordering characteristic of amyloid aggregates, i.e. linear propagation of local hydrophobicity peaks and troughs rather than a monocentric hydrophobic core (typically present in globular proteins). On this basis, the authors propose a hypothetical arrangement of V domains which leads to formation of an amyloid. Structural similarities between V domains in BJ proteins and other types of amyloid aggregates enable the authors to study the specific mechanism of Congo red complexation by amyloids.

The proposed Congo red complexation mechanism builds upon the authors' previous experience with bioinformatics tools. The subject should be of interest to researchers specializing in protein folding studies and misfolding diseases.

Keywords Hydrophobicity • Bence-Jones proteins • V domain of IgG • Amyloid • Force field • Congo red • Immunoglobulins • Amyloidgenesis • Supramolecular ligand • Misfolding

M. Banach (✉) • I. Roterman
Department of Bioinformatics and Telemedicine, Jagiellonian University – Medical College, Łazarza 16, 31-530, Krakow, Poland
e-mail: mateusz.banach@uj.edu.pl; myroterm@cyf-kr.edu.pl

B. Kalinowska
Faculty of Physics, Astronomy and Applied Computer Science, Jagiellonian University, Łojasiewicza 11, 30-348, Krakow, Poland
e-mail: malijka@gmail.com

L. Konieczny
Chair of Medical Biochemistry, Jagiellonian University – Medical College, Kopernika 7, 31-034 Krakow, Poland
e-mail: mbkoniec@cyf-kr.edu.pl

© The Author(s) 2018

I. Roterman, L. Konieczny (eds.), *Self-Assembled Molecules – New Kind of Protein Ligands*, https://doi.org/10.1007/978-3-319-65639-7_5

5.1 Characteristics of Immunoglobulin Domains

Complexation of supramolecular Congo red (CR) requires considerable relaxation of the target protein's native form. Immunoglobulin domains are known for their structural instabilities (amyloidogenic properties), and also for their ability to eagerly bind CR. In line with basic biochemical knowledge, the stability of proteins is determined by two factors: disulfide bonds and hydrophobic core.

Immunoglobulin domains are β-sandwiches composed of two distinct fragments, referred to as the upper core and the lower core respectively. Each domain includes one centrally placed disulfide bond linking both cores. While the stabilizing influence of this bond allows the domain to persist in its native form, individual domains vary greatly with respect to the structure of their hydrophobic cores [1]. This diversity is evidenced by structural analysis based on the fuzzy oil drop (FOD) model. FOD is an extension of the "oil drop" hydrophobicity distribution model, which introduced a binary distinction between the outer (hydrophilic) and inner (hydrophobic) layers [2]. The name of the model alludes to the notion of an oil drop immersed in water – the hydrophobic substance attempts to minimize its contact surface, becoming spherical in the process. Similarly, the protein folding process results in internalization of hydrophobic residues and exposure of hydrophilic residues on the surface of a globular capsule [3].

As already remarked, the fuzzy oil drop model extends the binary oil drop paradigm by introducing a continuous gradient of hydrophobicity between the core and the surface. This gradient is mathematically expressed by a 3D Gaussian, which peaks at the center of the molecule and then gradually decreases, reaching near-zero values at a distance of 3σ (where σ is the coefficient of the Gaussian).

In this chapter we will apply the fuzzy oil drop model in the analysis of immunoglobulin light chain domains exemplified by Bence-Jones proteins [4].

5.2 Target Proteins

Table 5.1 provides a summary of proteins selected for analysis. They are collectively referred to as Bence-Jones (BJ) proteins [4] and are exclusively of human origin. Each protein is a homodimer comprised by two identical IgG light chains.

The sole exception is 2Q1E, where two pairs of V domain dimers have been identified in the crystal structure. The reference protein is the Fab fragment of human immunoglobulin G (4PUB), consisting of the light chain (L) and the heavy chain (H).

The fuzzy oil drop has been applied in the analysis of the following structural units: both chains in complex; paired domains – V(L)-V(L) and C(L)-C(L), each domain individually and for 4PUB the complexes of domains V(L)–V(H) and C(L)-C(H). We have also assessed the status of each domain as a structural subunit of the complete dimer.

Table 5.1 Proteins selected for analysis

Protein – ID PDB	Dimer	Chain class	References
1B6D	L-L	KAPPA	[5]
1BJM	L-L	LAMBDA	[6]
1DCL	L-L	LAMBDA	[7]
1LIL	L-L	LAMBDA	[8]
2OLD	L-L		[9]
2OMB	L-L		[9]
2OMN	L-L		[9]
2Q1E	L-L-TETRAMER	KAPPA	[10]
3BJL	L-L	LAMBDA	[6]
4PUB	H-L		[11]

5.3 The Fuzzy Oil Drop Model

The fuzzy oil drop model has been extensively described in numerous publications [3, 12, 13]. The description presented below should therefore be regarded only as a brief introduction.

The traditional notion of a "hydrophobic core" refers to a concentration of hydrophobic residues at the geometric center of the protein, along with exposure of hydrophilic residues on its surface. When dealing with globular proteins, this configuration can be described with a 3D Gaussian, where the origin of the coordinate system coincides with the geometric center of the molecule and a separate σ coefficient is defined for each principal axis, delineating an ellipsoid capsule. The molecule itself should be oriented in such a way as to align its orthogonal dimensions (longest diagonals) with axes of the coordinate system. Values of σ_x, σ_y and σ_z are calculated as 1/3 of the separation between the origin of the system and the position of the most distant atom along each axis. This is schematically depicted in Fig. 5.1.

In accordance with the three-sigma rule, over 99% of the total volume of the Gaussian is captured by applying a cutoff distance of 3σ in each principal direction. The value of the Gaussian at any point within this ellipsoid capsule is interpreted as local theoretical hydrophobicity (also referred to as the "idealized" distribution).

The theoretical hydrophobicity distribution should be confronted with the observed distribution, which depends on local interactions between each residue and its neighbors. These calculations are based on the positions of the so-called effective atoms (averaged-out positions of all atoms comprising a given residue) and the intrinsic hydrophobicity of each amino acid. Each effective atom collects interactions with its neighbors, with a cutoff distance of 9 Å.

Theoretical (T) hydrophobicity is expressed by the following formulae:

$$\tilde{H}t_j = \frac{1}{\tilde{H}t_{sum}}\exp\left(\frac{-\left(x_j - \bar{x}\right)^2}{2\sigma_x^2}\right)\exp\left(\frac{-\left(y_j - \bar{y}\right)^2}{2\sigma_y^2}\right)\exp\left(\frac{-\left(z_j - \bar{z}\right)^2}{2\sigma_z^2}\right),$$

$\tilde{H}t_j$ is the theoretical hydrophobicity density (hence the t designation) at the jth point in the protein body. $\bar{x}, \bar{y}, \bar{z}$ correspond to the peak of the Gaussian in each of the three principal directions, while σ_x, σ_y, σ_z denote the range of arguments for each coordinate system axis. These coefficients are selected in such a way that 99% of the Gaussian's integral is confined to a range of $\bar{x} \pm 3\sigma$. Values of the distribution can be assumed to equal 0 beyond this range.

If the molecule is placed inside a capsule whose dimensions are given by $\bar{x} \pm 3\sigma_x, \bar{y} \pm 3\sigma_y, \bar{z} \pm 3\sigma_z$ then the values of the corresponding Gaussian represent the idealized hydrophobicity density distribution for the target protein. If $\sigma_x = \sigma_y = \sigma_z$, the capsule is perfectly spherical; otherwise it is an ellipsoid. The Gaussian yields hydrophobicity density values at arbitrary points in the protein body – for example at points which correspond to the placement of effective atoms (one per side chain). Ht_j is the hydrophobicity density determined for the jth amino acid while x, y and z indicate the placement of its corresponding effective atom.

The denominator of $\dfrac{1}{\tilde{H}t_{sum}}$ expresses the aggregate sum of all values given by the Gaussian for each amino acid making up the protein. This enables normalization of the distribution since $\tilde{H}t_j$ will always be equal to 1.0.

$\tilde{H}t_j$ values reflect the expected hydrophobicity density which should correspond to each amino acid in order for the hydrophobic core to match theoretical predictions with perfect accuracy, with all hydrophobic residues internalized and all hydrophilic residues exposed on the surface. The closer to the surface the lower the expected hydrophobicity density.

The position of each (jth) residue is represented effective atom localized at the geometric center of all atoms belonging to side chain of the residue under consideration (including Cα in the case of Gly). Protein encapsulation is presented in Fig. 5.1.

On the other hand, the actual distribution of hydrophobicity density observed (O) in a protein molecule depends on inter-chain interactions, which, in turn, depend on the intrinsic hydrophobicity of each amino acid. Intrinsic hydrophobicity can be determined by experimental studies or theoretical reasoning – our work bases on the scale published in [12] while the force of hydrophobic interactions has been calculated using algorithms proposed in [14]. For each amino acid j (or, more accurately, for each effective atom) the sum of interactions with its neighbors is computed and subsequently normalized by dividing it by the number of elementary interactions:

$$\tilde{H}o_j = \frac{1}{\tilde{H}o_{sum}} \sum_{N}^{i=1} \left(H_i^r + H_j^r \right) \begin{cases} \left[1 - \dfrac{1}{2} \left(7 \left(\dfrac{r_{ij}}{c} \right)^2 - 9 \left(\dfrac{r_{ij}}{c} \right)^4 + 5 \left(\dfrac{r_{ij}}{c} \right)^6 - \left(\dfrac{r_{ij}}{c} \right)^8 \right) \right] & \text{for } r_{ij} \leq c \\ 0 \text{ for } r_{ij} > c \end{cases}$$

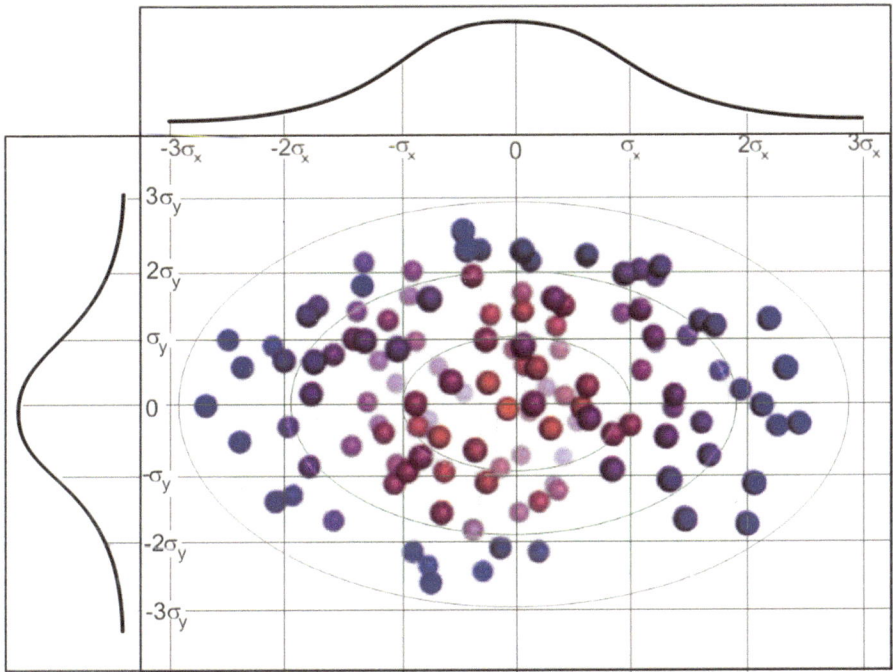

Fig. 5.1 Protein molecule placed in an ellipsoid capsule (only two dimensions are presented to preserve clarity). The adjacent one-dimensional plots present values of the Gaussian along each axis and highlight the role of σ coefficients. Color coding expresses the transition between hydrophilic (*blue*) and hydrophobic (*red*) residues. An intermediate layer is present between the core and the surface

N is the number of amino acids in the protein, $\tilde{H}_i^r$ expresses the hydrophobicity parameter of the i-th residue while r_{ij} expresses the distance between two interacting residues (jth effective atom and ith effective atom). c expresses the cutoff distance for hydrophobic interactions, which is taken as 9.0 Å (following [14]). The $\tilde{H}o_{sum}$ coefficient, representing the aggregate sum of all components, is needed to normalize the distribution which, in turn, enables meaningful comparisons between the observed and theoretical hydrophobicity density distributions.

Quantitative analysis of the differences between expected (T) and observed (O) distributions is enabled by the Kullback-Leibler entropy formula [15]:

$$D_{KL}\left(p|p^0\right) = \sum_{N}^{i=1} p_i \log_2\left(p_i / p_i^0\right)$$

The value of D_{KL} expresses the distance between the observed (p) and target (p_0) distributions, the latter of which is given by the 3D Gaussian (T). The observed distribution is referred to as O.

For the sake of simplicity, we introduce the following notation:

$$O/T = \sum_{N}^{i=1} O_i \log_2 O_i / T_i$$

Since D_{KL} is a measure of entropy its interpretation requires a reference value. In order to facilitate meaningful comparisons, we introduce another boundary distribution (referred to as "unified" or R) which corresponds to a situation where each effective atoms represents the same hydrophobicity density ($1/N$, where N is the number of residues in the chain). In this type of distribution hydrophobicity density is not concentrated at any point in the protein body.

$$O/R = \sum_{N}^{i=1} O_i \log_2 O_i / R_i$$

Comparing O|T and O|R tells us whether the given protein (O) more closely approximates the theoretical (T) or unified (R) distribution. Proteins for which O|T > O|R are regarded as lacking a prominent hydrophobic core. To further simplify matters we introduce the following relative distance (RD) criterion:

$$RD = \frac{O/T}{O/T + O/R}$$

Here, RD < 0.5 indicates the presence of a hydrophobic core.

Figure 5.2 presents a graphical representation of RD values, restricted (for simplicity) to a one dimensional form.

D_{KL} (as well as O|T, O|R and RD) may be calculated for specific structural units (complex, single molecule, single chain, selected domain). In such cases the bounding ellipsoid is restricted to the selected fragment of the protein. It is also possible to determine the status of polypeptide chain fragments within the context of a given ellipsoid. This procedure requires prior normalization of O_i, T_i and R_i values belonging to the analyzed fragment.

The procedure described above will be consistently applied in the analysis presented in this chapter. The status of selected polypeptide chain fragment will be studied to evaluate their participation in forming a hydrophobic core. In particular, secondary folds which satisfy RD < 0.5 are thought to contribute to the molecule-wide hydrophobic core. When the opposite is true (i.e. RD > 0.5), the given fragment can be considered unstable. It appears that fragments which exhibit higher-than-expected hydrophobicity may, when exposed on the surface, be engaged in protein complexation (forming parts of the interface).

Calculations concerning fragments of the polypeptide chain requires prior normalization of T_i, O_i and R_i values belonging to the selected fragment. The results tell us whether the given fragment contributes to the molecule-wide hydrophobic core.

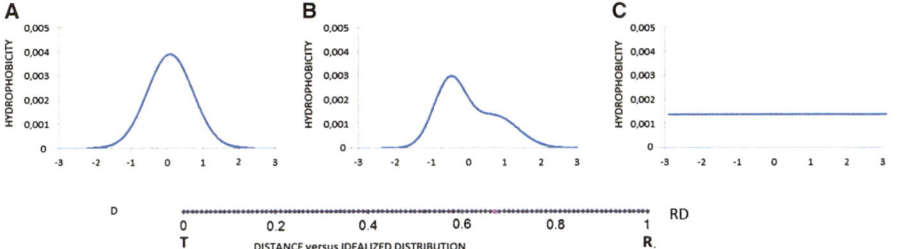

Fig. 5.2 Graphical representation of fuzzy oil drop model parameters, reduced to a single dimension. (**A**) theorized Gaussian distribution (T). (**B**) actual hydrophobicity distribution in the protein under consideration. (**C**) uniform distribution (R). (**D**) The RD parameter (equal to 0.656) marked on the horizontal axis as a *pink dot*. According to the fuzzy oil drop model this protein does not contain a well-defined hydrophobic core

Fragments can be selected for analysis on the basis of their involvement in particular secondary structures [1], supersecondary structures [16], interface areas [17], intrinsically distorted fragments [18], chameleon fragments [19] or other types of structures, depending on the research problem at hand.

A summary of sample proteins, visualizing the varied status of their hydrophobic cores, is provided in Fig. 5.3. Titin (Fig. 5.3A) is a protein which includes an immunoglobulin-like fold, exhibiting very good agreement between the theoretical and the observed hydrophobicity distribution. In this case hydrophobicity is concentrated near the center of the protein, with a hydrophilic layer present on the surface, optimizing the protein's contact with water. Such high agreement between T and O enables titin to revert to its native conformation in the absence of external forces (note that titin is found in muscle tissue and subject to frequent stretching).

The second sample protein, visualized in Fig. 5.3B, is the H chain of the human immunoglobulin Fab fragment. In this case major discrepancies between the theoretical and observed hydrophobicity distribution are observed (RD = 0.584), indicating that no monocentrichydrophobic core is present and that the H chain as a whole is only marginally stable. Further stabilization is provided by two disulfide bonds present in the Fab fragment. Notably, immunoglobulin appears to require a flexible V domain in order to align itself with the antigen.

The third protein, transthyretin, is a known amyloid precursor (Fig. 5.3C). Major differences between the N-terminal and the C-terminal fragments are evident. The protein as a whole does not follow the theoretical distribution of hydrophobicity, although the N-terminal section is a far better match for the theoretical values than its C-terminal counterpart. In such cases, it is informative to compute RD values for specific fragments of the chain, revealing the degree of their participation in the protein's hydrophobic core.

The presented work focuses on Bence-Jones complexes formed by IgG light chains [4]. Detection of such proteins in urine may indicate multiple myeloma or Waldenström's macroglobulinemia. Large deposits of B-J proteins are also encountered in kidneys and may cause amyloidogenesis [20].

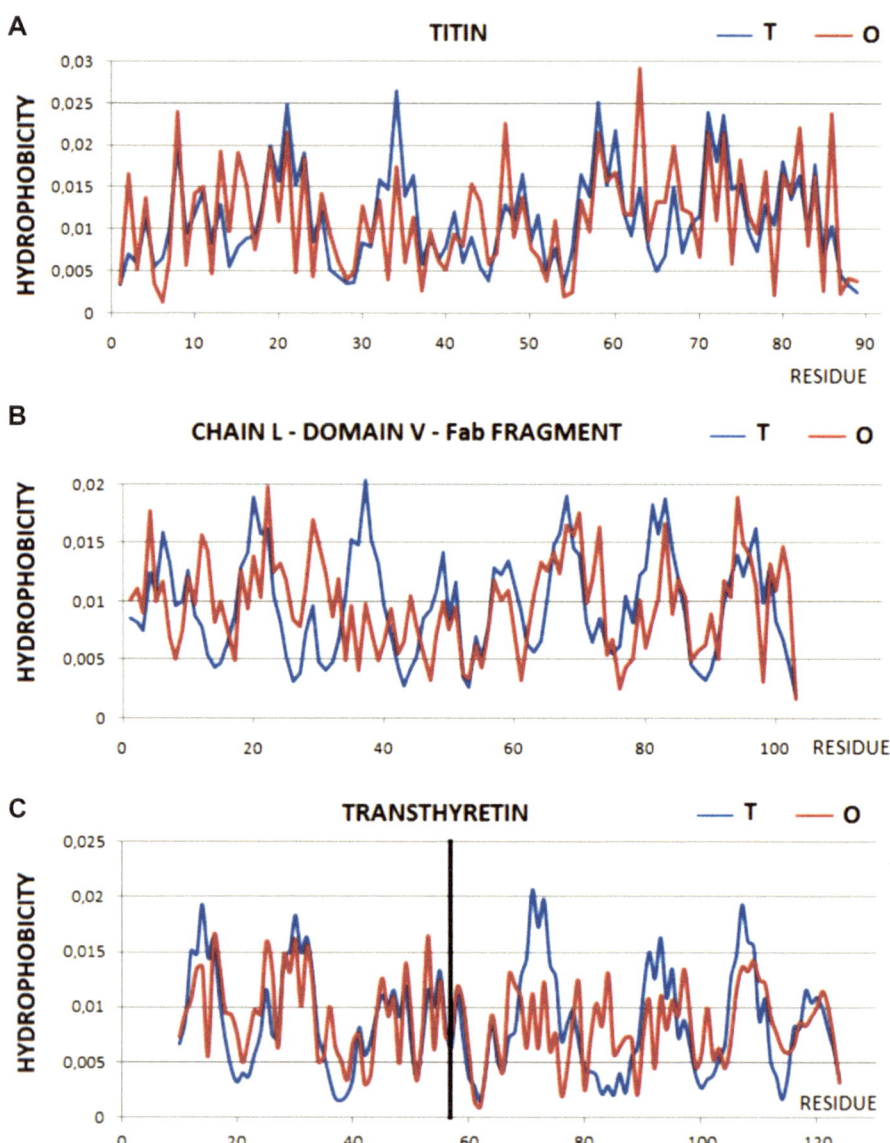

Fig. 5.3 Comparison of three proteins: (**A**) O matches T - immunoglobulin-like domain; titin; RD = 0.326. (**B**) O diverges from T – domain V chain L; human immunoglobulin Fab fragment; RD = 0.584. (**C**) O partially matches T; however areas of significant discordance are evident – transthyretin; RD = 0.562 (entire chain), 0.475 (accordant fragment at 10–57), 0.621 (discordant fragment at 51–124). The *vertical line marks* the boundary between both fragments

The reference complex is provided by the light/heavy chain dimer corresponding to the native form of immunoglobulin (4PUB).

5.4 Structure of Hydrophobic Core in B-J Proteins

The fuzzy oil drop model has been applied to identify the status of immunoglobulin domains in proteins referred to as Bence-Jones complexes. This choice of proteins is motivated by their specific properties, particularly their ability to quickly transition into amyloid forms (indicating structural instabilities) and their high affinity for supramolecular CR.

The fuzzy oil drop model reveals the status of the complete protein (complex), its individual domains as well as fragments of polypeptide chains – the presented analysis covers the status of the interface area and the N-terminal fragment which has been experimentally characterized as highly unstable [21].

5.4.1 Dimers of L-L Chains in Bence-Jones Proteins and of L-H Chains in the Fab Fragment

Table 5.2 presents a comparison of RD parameters describing full-chain dimers and V/C domains present in the complex. Analysis of results indicates that the full-chain dimer does not contain a shared hydrophobic core in the sense of the fuzzy oil drop model (with all corresponding RD values in excess of 0.5).

When analyzed as components of the complex, V domains exhibit high RD values, which suggests that they lack prominent monocentrichydrophobic cores. Two exceptions to this rule are 1LIL and 2Q1E. In the latter case, the discrepancy is due to altered composition of the V domain tetramer crystals, where each unit cell comprises two dimers with two domains per dimer. For this reason, 2Q1E will be frequently seen as an outlier in further analysis.

C domains analyzed in the context of the dimer exhibit good agreement with the theoretical hydrophobic core structure.

Interesting properties are revealed for the light/heavy chain complex in the Fab IgG fragment, where each domain is discordant in the context of the complex-wide hydrophobic core.

5.4.2 V(L)-V(L), C(L)-C(L), V(H)-V(L) and C(H)-C(L) Dimers

Since immunoglobulin chains consist of clearly distinguished paired domains, it is interesting to study the status of V(L)-V(L) and C(L)-C(L) dimers. Table 5.3 provides the corresponding quantitative characterization. In the case of the Fab fragment, we have analyzed its V(L)-V(H) and C(L)-C(H) dimers, which should be regarded as a reference.

Table 5.2 RD values describing individual domains of light chain dimers in Bence-Jones proteins

Protein – ID PDB	Dimer	Chain/domain V	RD	Chain/domain C	RD
1B6D	**0.778**	A-V (1-107)	**0.541**	A-C (108-211)	0.470
		B-V (1-107)	**0.566**	B-C (108-211)	0.417
1BJM	**0.718**	A-V (2-111)	**0.558**	A-C (112-212)	0.383
		B-V (2-111)	**0.590**	B-C (112-212)	0.331
1DCL	**0.752**	A-V (1-111)	**0.588**	A-C (112-212)	0.344
		B-V (1-111)	**0.567**	B-C (112-212)	0.343
1LIL	**0.733**	A-V (2-107)	0.473	A-C (108-211)	0.347
		B-V (2-107)	0.485	B-C (108-211)	0.336
2OLD	**0.734**	A-V (2-112)	**0.588**	A-C (113-213)	0.376
		B-V (2-112)	**0.591**	B-C (113-213)	0.310
2OMB	**0.803**	A-V (2-112)	**0.597**	A-C (113-213)	0.420
		B-V (2-112)	**0.589**	B-C (113-213)	0.303
		C-V (2-112)	**0.602**	C-C (113-213)	0.351
		D-V (2-112)	**0.600**	D-C (113-213)	0.347
2OMN	**0.750**	A-V (2-112)	**0.592**	A-C (113-213)	0.390
		B-V (2-112)	**0.581**	B-C (113-213)	0.300
2Q1E	**0.786**	A-V	0.492	A-C	0.492
		B-V	0.507	B-C	**0.507**
		C-V	0.489	C-C	0.489
		D-V	0.464	D-C	0.464
3BJL	**0.712**	A-V (2-111)	**0.583**	A-C (112-212)	0.375
		B-V (2-111)	**0.586**	B-C (112-212)	0.312
4PUB	**0.747**	H-V (2-122)	**0.754**	H-C (123-223)	**0.663**
		L-V (1-107)	**0.818**	L-C (108-212)	**0.728**

The reference protein (4PUB) is the Fab fragment of human immunoglobulin, composed of light (L) and heavy (H) chains

Table 5.3 RD values for V and C domain dimers

Protein	V-V dimer			C-C dimer		
	Complete	No P-P	P-P	Complete	No P-P	P-P
1B6D	**0.752**	**0.573**	**0.630**	**0.538**	**0.508**	**0.505**
1BJM	**0.683**	**0.648**	0.492	**0.529**	0.492	0.317
1DCL	**0.751**	**0.732**	**0.505**	**0.540**	**0.513**	0.291
1LIL	**0.695**	**0.663**	**0.643**	**0.502**	0.445	0.245
2OLD	**0.714**	**0.697**	**0.534**	**0.546**	**0.525**	0.300
2OMB	**0.755**	**0.731**	**0.745**	**0.542**	**0.521**	**0.535**
2OMN	**0.750**	**0.728**	**0.596**	**0.532**	**0.520**	0.283
2Q1E	**0.753**	**0.722**	0.475			
3BJL	**0.655**	**0.630**	0.385	**0.561**	**0.528**	0.291
4PUB	**0.649**	**0.603**	0.450	**0.531**	**0.515**	0.373

The "no P-P" columns characterize the status of domains following elimination of residues involved in P-P interactions, while the "P-P" columns presents the interface fragments. Values listed in boldface represent discordance

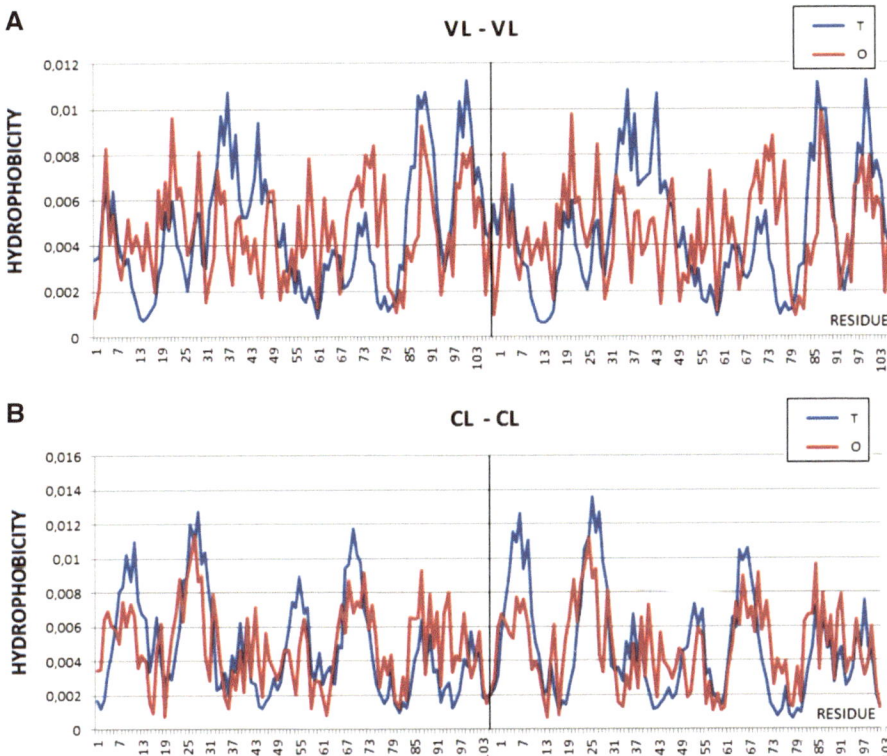

Fig. 5.4 Hydrophobicity distributions in protein domains: (**A**) V(L)-V(L), (**B**) C(L)-C(L). The divergence between T and O is greater in V domains than in the C(L)-C(L) dimer

Figure 5.4 presents the theoretical and observed distribution of hydrophobicity in 1B6D (B-J protein). Of note is the significant deviation corresponding to the V domain, with only the C-terminal complex consistent with the theoretical distribution. Comparison of V domain profiles with their C domain counterparts reveals greater stability of the C dimer (at least from the point of view of its hydrophobic core structure).

In all listed proteins, full V-V and C-C dimers deviate from the expected monocentric distributions of hydrophobicity. Elimination of residues involved in interdomain interactions produces a reduction in RD values, showing that the interface zone disrupts the structure of the hydrophobic core in each domain.

On the other hand, when analyzing the distribution of hydrophobicity in the interface itself, it turns out that a significant majority of C-C dimers match the theoretical distribution, and the same is true for the Fab fragment. Evidently, the distribution of hydrophobicity in interface residues corresponds to FOD predictions (except in 1B6D and 2OMB).

In general, the structure of C-C dimers may be interpreted as relatively stable, whereas V-V dimers are characterized by low stability.

5.4.3   Individual Domains

This part of the presentation focuses on individual domains regarded as standalone structural units. For each domain, a separate 3D Gaussian is plotted and the corresponding RD values calculated. Results are listed in Table 5.4.

If we base our analysis on the textbook definition of a domain (i.e. a distinct structural unit which folds on its own), the FOD properties of standalone domains should reveal their intrinsic structural stability. Analysis of results presented in Table 5.4 suggests significant differences between V and C domains. The former are generally discordant (lack hydrophobic cores), while in the latter case a prominent hydrophobic core is present for each analyzed B-J protein, with the sole exception being 1LIL. Eliminating residues involved in inter-domain interactions bring the status of V domains in line with the theoretical model – we may therefore conclude that structural instabilities are primarily due to the presence of an inter-domain interface. This theory is corroborated by the poor agreement between T and O in the interface itself.

Contrasting properties are observed in the V(L) and V(H) domains comprising the IgG Fab fragment, with both units conforming to the model. The status of Fab V(L) is similar to that of its B-J counterparts.

The observed discrepancies between the status of V and C domains in both chains (L and H) provide important clues regarding amyloidogenesis. This phenomenon is more frequently observed in B-J proteins, although L-H dimers are not immune from it. Notably, amyloidogenesis tends to involve V domains rather than C domains [22].

As already remarked, the peculiar status of 2Q1E is due to differences in its crystal structure, with a marked decrease in the quantity of residues involved in inter-domain interactions (16 compared to 29–38 in other dimers). Regarding 1LIL, its dimerization properties differ from other proteins in the study set due to structural differences in the interface zone.

5.5   Role of the N-Terminal Fragment in V Domains

The N-terminal fragment of the light chain V domain has been identified as highly unstable on the basis of experimental results [21]. This conclusion is supported by molecular dynamics simulations involving B-J proteins.

Eliminating the N-terminal fragment results in a significant decrease in RD values, proving that the fragment contributes to destabilization of the domain and disrupts its hydrophobic core (Table 5.5).

The disagreement between O and T distribution as observed in N-terminal fragment is visualized in Fig. 5.5. The status of observed hydrophobicity distribution of position 5 and fragment 11–14 can be even treated as opposite one versus the expected hydrophobicity. RD value for this fragment is equal to 0.604.

Table 5.4 RD values for individual domains comprising the presented proteins

| Protein | RD Values | | No P-P interaction | | P-P interface | |
| | Domain | | | | | |
	Domain V	Domain C	Domain V	Domain C	Domain V	Domain C
1B6D	**0.541/0.566**	0.470/0.417	0.195/0.183	0.471/0.414	**0.602/0.626**	0.428/0.481
1BJM	**0.558/0.590**	0.383/0.331	0.217/0.184	0.383/0.352	**0.664/0.730**	0.361/0.159
1DCL	**0.588/0.567**	0.344/0.343	0.178/0.173	0.345/0.348	**0.506/0.522**	0.287/0.234
1L1L	0.474/0.485	0.347/0.336	0.239/0.242	0.344/0.348	**0.539/0.693**	0.284/0.210
2OLD	**0.588/0.591**	0.376/0.310	0.159/0.151	0.386/0.317	**0.645/0.613**	0.145/0.177
2OMB	**0.597/0.589**	0.420/0.303	0.153/0.178	0.416/0.308	**0.664/0.731**	0.396/0.272
2OMN	**0.592/0.581**	0.390/0.301	0.154/0.164	0.389/0.313	**0.603/0.646**	0.224/0.173
2Q1E	0.492/0.507		0.473/0.485		**0.834/0.872**	
	0.490/0.465		0.473/0.420		**0.820/0.738**	
3BJL	**0.583/0.586**	0.375/0.312	0.189/0.193	0.374/0.327	**0.648/0.784**	0.257/0.117
4PUB (H/L)	0.473/0.460	**0.595/0.480**	**0.594/0.465**	**0.594/0.496**	0.393/0.384	**0.568/0.323**

Reference data is provided for 4PUB (Fab fragment). Calculations are performed for each domain separately

Table 5.5 RD values for complete V domains (central column) and following elimination of the N-terminal fragments (right column)

Protein	Domain V	
	Complete	No N-Terminal Fragment
1B6D	**0.541/0.566**	0.211/0.198
1BJM	**0.558/0.590**	0.219/0.189
1DCL	**0.588/0.567**	0.208/0.204
1LIL	0.474/0.485	0.244/0.241
2OLD	**0.588/0.591**	0.168/0.168
2OMB	**0.597/0.589**	0.166/0.175
2OMN	**0.592/0.581**	0.175/0.177
2Q1E	0.492/**0.507**	0.432/0.442
	0.490/0.465	0.433/0.420
3BJL	**0.583/0.586**	0.196/0.190
4PUB (H/L)	0.460	0.453

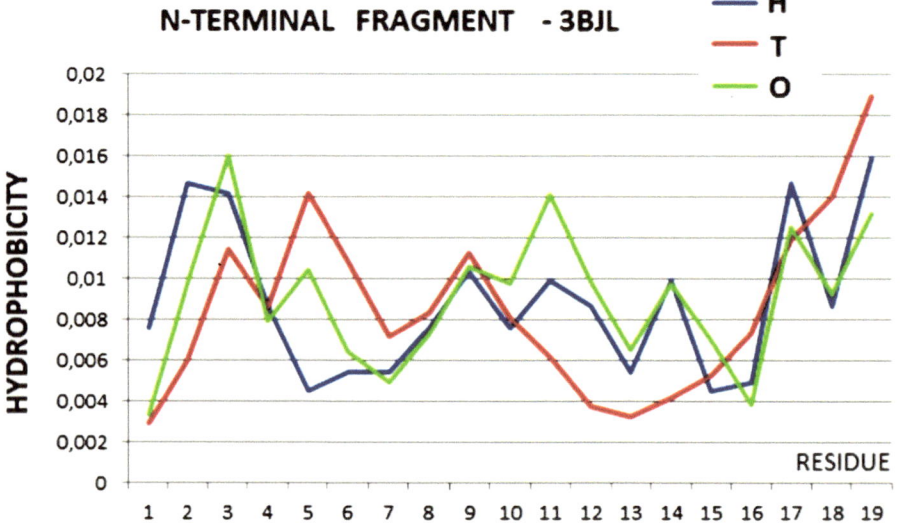

Fig. 5.5 The N-terminal fragment – profiles of T, O and H distribution

The N-terminal section in the BJ dimer occupies an exposed position, facing the environment. This renders it susceptible to structural changes – it may become uncoiled, freeing itself from the influence of the protein (in the sense of the FOD model) [21]. Figure 5.6 depicts this situation distinguishing the N-terminal fragment and fragments engaged in interface generation as the region of lower stability due to lover engagement in hydrophobic core generation.

According to fuzzy oil drop model – fragments of the polypeptide chain which do not contribute to the molecule-wide hydrophobic core are regarded as potentially unstable and susceptible to structural changes.

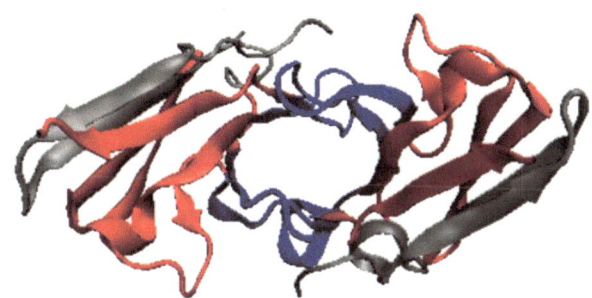

Fig. 5.6 3D visualization of the V domain in a B-J protein (3BJL), showing fragments which diverge from the theoretical hydrophobicity distribution: *gray* – N-terminal fragment; *dark blue* – interface area. VMD program was used to draw the picture [23]

5.6 Hypothetical Amyloidogenesis Mechanism Affecting the V Domain in B-J Proteins

Applying the fuzzy oil drop model to amyloid structures points to linear propagation of local hydrophobicity distributions along the long axis of the fibril. The model proposed in [24] stipulates that β-structural fragments perpendicular to the fibril's axis exhibit the following properties:

1. discordant distribution of hydrophobicity vs. theoretical values (there is no monocentric core which would ensure formation of a globular protein);
2. interspersed peaks and troughs of hydrophobicity observed along the unit β-structural fragment;
3. if the amyloid is composed of identical polypeptides, the identical local distribution is repeated for each unit peptide, with linear propagation of hydrophobicity peaks/troughs along the long fibril axis. Similar linear propagation is observed for β-structural fragments with varying sequences as long as their overall profiles remain similar.

All these conditions can be observed in the 2MVX amyloid [25].

Figure 5.7 summarizes the differences between the observed and theoretical distribution (the latter of which would ensure the formation of a centralized hydrophobic core). Each fragment (including β-structural ones) is sequentially identical to all other folds comprising the amyloid; thus the presented distribution of hydrophobicity can be repeated linearly, along with the corresponding propagation of local peaks and troughs (Fig. 5.8).

If linear propagation of local hydrophobicity profiles is taken as a criterion for identifying amyloid forms, then the presence of such arrangement should be regarded as a seed for further amyloid aggregation. Under certain conditions (e.g. shaking) linear propagation effectively "outcompetes" the standard folding process which usually produces monocentrichydrophobic cores (see titin profile on Fig. 5.3A). The folding process follows the intrinsic hydrophobicity rather than generates the common unicentric construction of hydrophobic core. The resulting amyloidogenesis may affect either the entire protein (domain), or enable multiple proteins to cluster together.

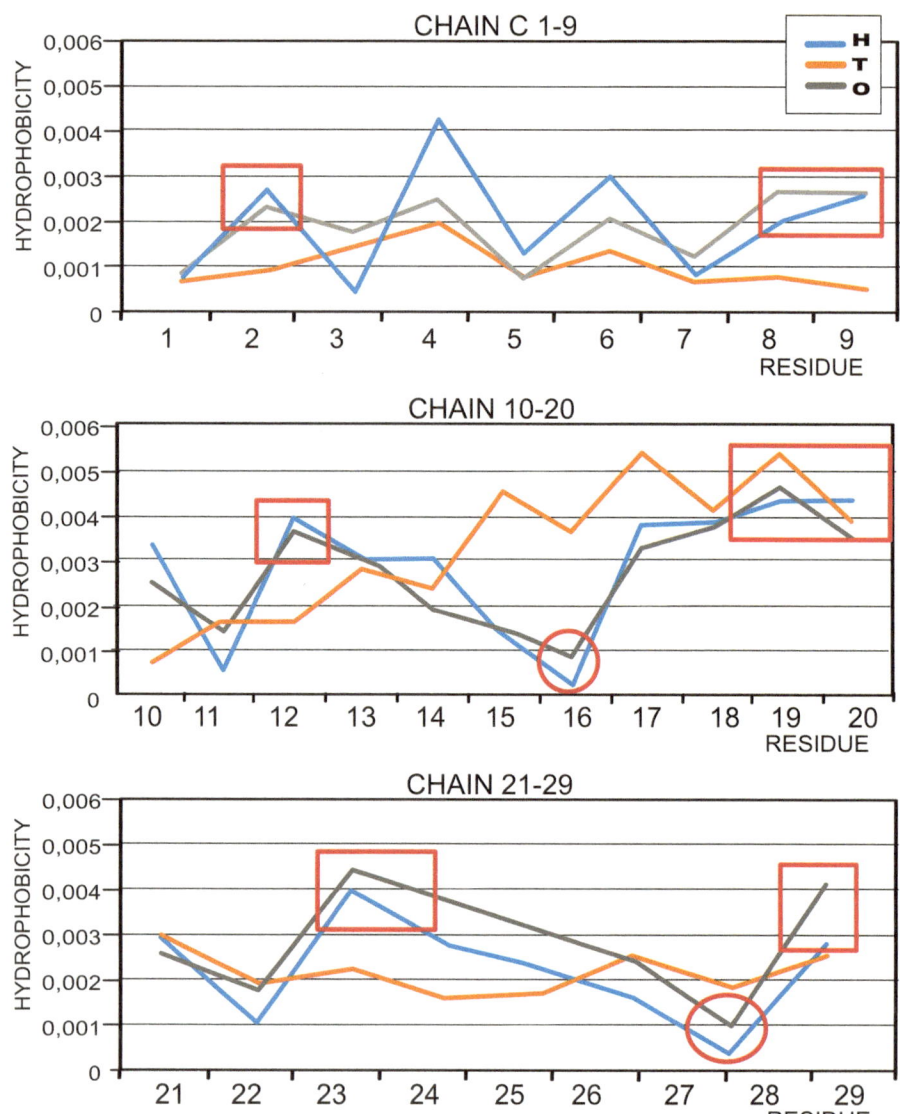

Fig. 5.7 Hydrophobicity distribution profiles in fragments as they appear in 2MVX: H – intrinsic hydrophobicity corresponding to each amino acid; T – theoretical (expected) hydrophobicity given by the FOD model; O – actual (observed) hydrophobicity resulting from inter-residue interactions. *Red* frames mark local peaks, while *red circles* denote local troughs, both of which represent deviations from the theoretical model. Selected fragments are listed above each chart. The profiles shown are identical along the long axis of the amyloidfibril due to identical sequence of peptides generating the fibril

Fig. 5.8 3D presentation of linear propagation of hydrophobicity peaks (*red*) and troughs (*blue*). This arrangement is markedly different from the theoretical distribution, where hydrophobic residues are expected to cluster at the center of the molecule while hydrophilic residues remain exposed on its surface (as shown in Fig.5.7). VMD program was used to draw the picture [23]

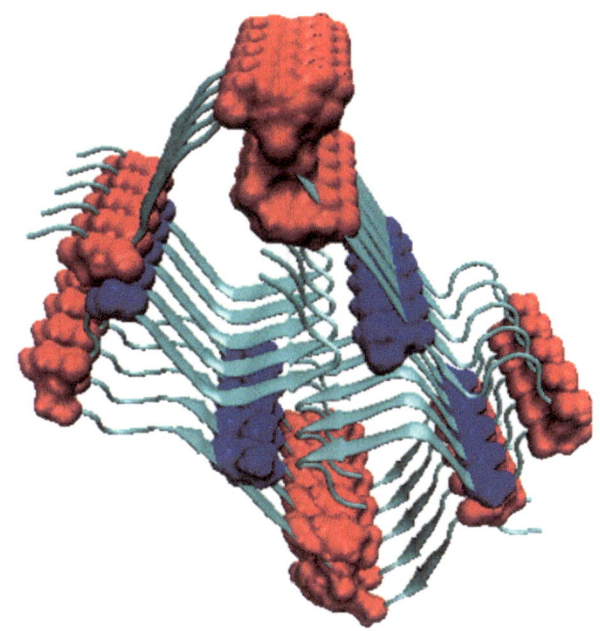

The presented distribution is evident in the V domain of the IgG light chain (crystal structure – 3BJL), and particularly its β-structural fragments at 86–90, 40–35, 45–50 and 56–51 (note the antiparallel arrangement). Figure 5.9 presents the distribution of hydrophobicity in each of these fragments.

As shown in Fig. 5.9, some β-structural fragments deviate from the theoretical distribution of hydrophobicity (the RD value for 35–40 equals to 0.676; for 53–60 fragment equals to 0.593). Those which appear to demonstrate the accordant distribution (fragment 43–50 described by RD = 0.463; fragment 86–90 by the RD value = 0.422) in selected positions represent the hydrophobicity level locally different in respect to the expected one. What is more, the specific local profiles of adjacent β-structural fragments enable linear propagation. Figure 5.10 provides a 3D depiction of this phenomenon. Local minima are bracketed by local peaks, all of which propagate linearly and diverge from the theoretical distribution (Fig. 5.10). We can also observe a clear correspondence between the observed distribution of hydrophobicity and the intrinsic hydrophobicity of each participating amino acid. It means no tendency to create the common unicentrichydrophobic core is observed. The residues in polypeptide chain accept the conformation following its intrinsic hydrophobicity.

The 3D visualization shown on Fig. 5.10 corresponds to the profiles shown in Fig. 5.9.

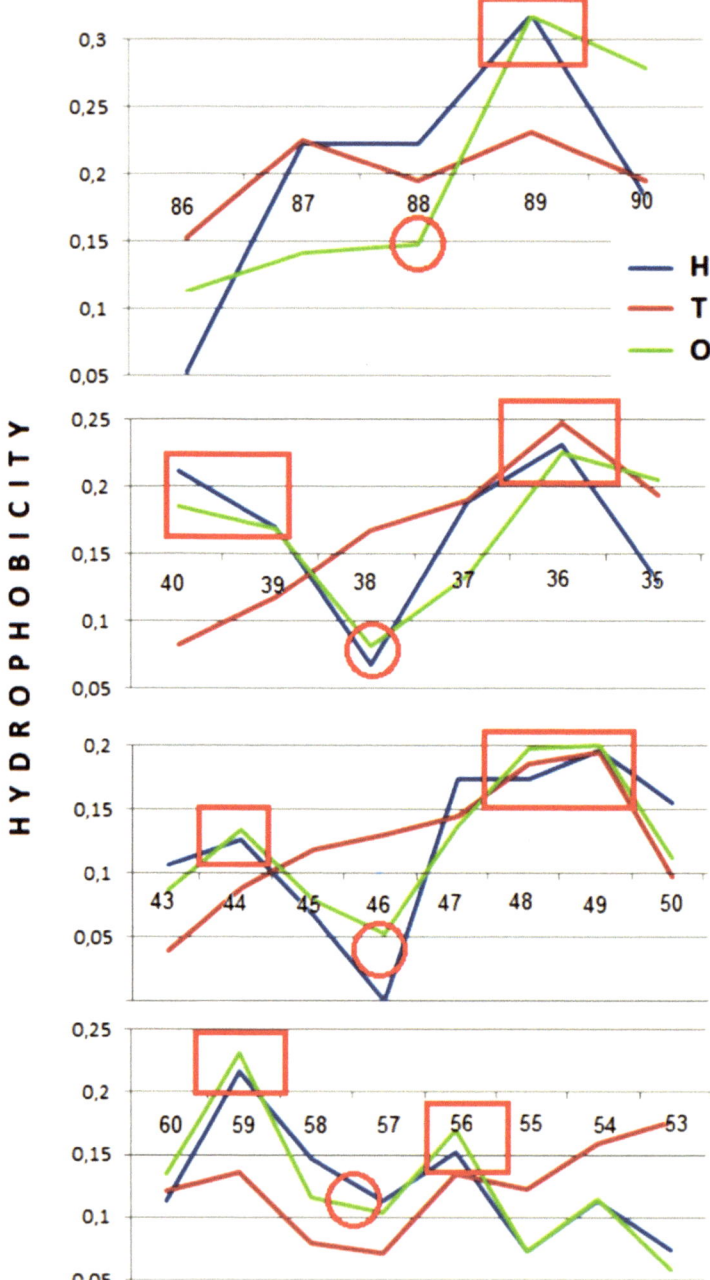

Fig. 5.9 Hydrophobicity distributions in the V domain (H – intrinsic; T – theoretical; O – observed), plotted for fragments which exhibit linear propagation of local troughs (*red circles*) and peaks (*red frames*). The order of residues in fragments accordant to anti-parallel orientation of β-structural fragments

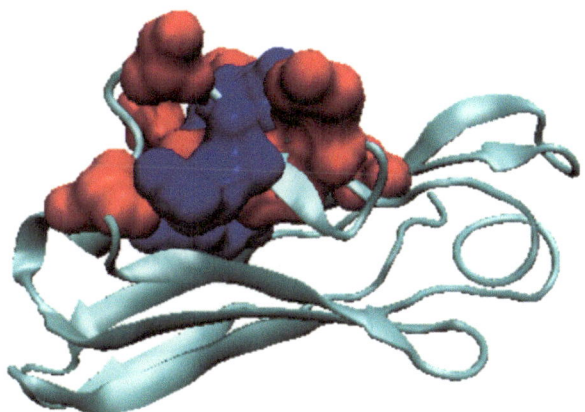

Fig. 5.10 3D presentation of the V domain. *Red* – propagation of hydrophobicity; *blue* – propagation of hydrophilicity. The hydrophilic band in the middle of the β-structural fragment, bracketed by local peaks, provides a seed for linear propagation. The resulting distribution is a poor match for theoretical values (which predict exposure of hydrophilicity on the protein surface). VMD program was used to draw the picture [23]

5.7 Complexation of Congo Red

According to the experiences with CR binding two models can be distinguished:

1. Supramolecular CR may serve as a ligand for any protein, as long as the protein contains a suitable docking cavity. The ligand micelle wedges itself between two adjacent β-structural fragments, as described in [26, 27]. Under these conditions the target protein usually retains its monocentrichydrophobic core, along with any local deviations associated with natural ligand binding capabilities [28]. The supramolecular ligand may occupy the space vacated by a displaced loop, as indeed observed in the case of IgG V domain [21]. The potential ligand binding site can be recognized using FOD calculation as local hydrophobicity deficiency [28]. See also Fig. 2.4.

2. Supramolecular CR is known for its ability to bind to amyloids. An open question concerns the manner in which the dye attaches itself to amyloid aggregations – it can be suspected that the mechanism differs from complexation of individual proteins. In an amyloid, the putative monocentrichydrophobic core is replaced with a distribution of hydrophobicity which reflects the intrinsic properties of each participating residue. If two or more β-structural fragments exhibit similar hydrophobicity profiles, the likelihood of linear aggregation is increased. When the entire V domain converts to a form dominated by linear propagation of hydrophobicity, a multidomain fibril may emerge, as schematically illustrated in Fig. 5.11. It is notable that in such situations the ribbonlike dye micelle complements the linear form of the amyloid itself (which can also be treated as ribbonlike or cylindrical micelle). The orientation of the CR micelle (linear propagation

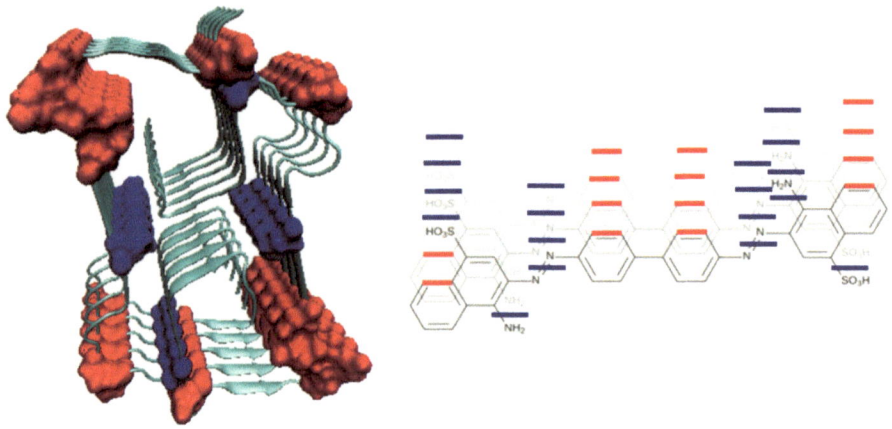

Fig. 5.11 Similarities – linear propagation – in the distribution of local hydrophobicity peaks (*red*) and troughs (*blue*) in an amyloid plaque (2MVX; a4 β-amyloid) and in the CR micelle. VMD program was used to draw the 3D picture [23] and BKChem to draw the CR formula [29]

of local hydrophobicity/hydrophilicity: aromatic rings vs. sulfonic and azo-groups) in respect to linear propagation of hydrophobicity peaks (or troughs) in amyloid fibrils (as shown in Fig. 5.8) appears to be compatible and able to align axially with each other. This observation is not invalidated by the presence of axial twists (as discussed in numerous publications), since in this respect the same property is shared by the amyloid fibril and the dye micelle making mutual adaptation still possible. The liquid crystal form of the CR micelle can align itself to a wide variety of linearized hydrophobicity distributions, which explains why the dye is capable of forming complexes with various amyloid fibrils – both protein-based and peptide-based.

5.8 Hypothetical Amyloidogenesis of V Domains

In summary, we can conclude that – at least according to the fuzzy oil drop model – stabilization of BJ dimers is mediated by C domain complexes. In contrast, the V domains appear quite unstable. RD values calculated for C-C complexes are only slightly above 0.5, while their V-V counterparts are much higher. When structural changes are expected in the dimer, the fuzzy oil drop model points to the V domain as the preferential location of such changes.

The N-terminal fragment adjacent to the interface zone (Fig. 5.5; Tables 5.4 and 5.5) is particularly prone to conformational changes resulting from its poor alignment with the theoretical hydrophobicity distribution. Such changes may provide the seed for amyloid transformation – although the issue is quite complex and requires further study [13, 24].

Figure 5.12 shows a hypothetical mechanism of multimolecular fibril generation. The status of β-structural fragments (as shown in Fig. 5.9) suggests the possible propagation of local maxima and local minima of hydrophobicity in contrast to expected distribution for these fragments. As it is shown in Fig. 5.12. the approach of two units representing similar characteristics is able to make possible propagation of linear hydrophobicity/hydrophilicity propagation. The red fragments on Fig. 5.12 are those shown in Fig. 5.9. According to 3D presentation the fragment 53–60 (distinguished as pink) is expected to fit its structure to the partner from the next unit (domain). Loose N-terminal fragments, devoid of hydrophobic stability (as confirmed by molecular dynamics simulations and experimental studies [21, 26]), may align with one another, creating a new β-interface especially due to its localization on the edge of the domain.

The crystal structure of the V-V dimer does not correspond to the actual conformation of these domains in an amyloidfibril. Certain structural changes are expected in the N-terminal fragment (Fig. 5.6), but also in the domain as a whole (see [13] for a discussion of potential changes expected for fibril formation). In light of this fact, it is difficult to speculate about the final structure of the V-V amyloid – although conformational rearrangements proposed for transthyretin [13], converting its crystal structure into an amyloid, appear equally possible in the IgG V domain (as seen in 3BJL).

The proposed supramolecular CR binding mechanism – one of many possible – is superficial in nature and does not require the dye to penetrate the amyloid. This explains why CR is able to adhere to amyloids formed by separate domains, as well as by identical β-structural fragments comprising a single domain. Due to the specific type of inter-chain interactions occurring in amyloids, intercalation of the dye is unlikely, and while CR may potentially dock in a suitable cavity (as proposed in

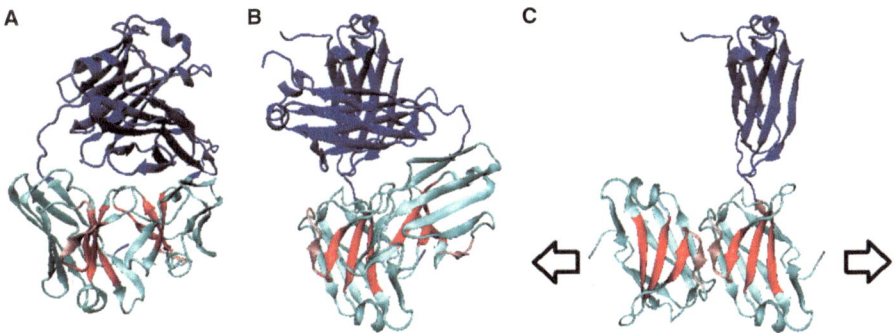

Fig. 5.12 3D presentation of possible interactions between V domains (**A** and **B**) different perspective. (**C**) Possible propagation of β-sheet in amyloid form. *Arrows* indicate possible directions of propagation. The same situation occurs in the complementary chain (second IgG light chain – V domain). *Red* – fragments consistent with the hydrophobicity distribution shown in Fig.5.8 and therefore capable of linear propagation. The *pink* fragment (53–60) requires conformational changes in order to adapt itself to linear propagation of hydrophobicity minima/maxima in β-structural fragments to the next adjacent unit. VMD program was used to draw the picture [23]

[27] see also Fig. 2.4), this mode of complexation is characteristic of individual proteins rather than amyloid fibrils. Finally, the supramolecular CR micelle may also wedge itself between parallel amyloid strains, as observed in 2MVX. Additionally both forms of supramolecular micelle (CR) as well as amyloid by itself are sensitive to external conditions reacting and adopting forms adequate to environmental factors [30].

The presented mechanism of the V domains transformation into the amyloidfibrils can be treated as possible one, obviously assuming that the proposed model of amyloidogenesis as the tendency to linear propagation of hydrophobic characteristics is acceptable. The detergent-like interaction of CR with amyloid described in this chapter is additionally supported by observation reported in [31].

Acknowledgements Work financially supported by Collegium Medicum – Jagiellonian University grant system – grant # K/ZDS/006363.

References

1. Banach M, Konieczny L, Roterman I (2014) The fuzzy oil drop model, based on hydrophobicity density distribution, generalizes the influence of water environment on protein structure and function. J Theor Biol 359:6–17
2. Kauzmann W (1959) Some factors in the interpretation of protein denaturation. Adv Protein Chem 14:1–63
3. Konieczny L, Brylinski M, Roterman I (2006) Gauss-function-Based model of hydrophobicity density in proteins. Silico Biol 6(1–2):15–22
4. Bernier GM, Putnam FW (1963) Monomer-dimer forms of Bence-Jones proteins. Nature 200:223–225
5. Roussel A, Spinelli S, Déret S, Navaza J, Aucouturier P, Cambillau C (1999) The structure of an entire noncovalent immunoglobulin kappa light-chain dimer (Bence-Jones protein) reveals a weak and unusual constant domains association. Eur J Biochem 260(1):192–199
6. Huang DB, Ainsworth CF, Stevens FJ, Schiffer M (1996) Three quaternary structures for a single protein. Proc Natl Acad Sci U S A 93(14):7017–7021
7. Ely KR, Herron JN, Harker M, Edmundson AB (1989) Three-dimensional structure of a light chain dimer crystallized in water. Conformational flexibility of a molecule in two crystal forms. J Mol Biol 210(3):601–615
8. Huang DB, Ainsworth C, Solomon A, Schiffer M (1996) Pitfalls of molecular replacement: the structure determination of an immunoglobulin light-chain dimer. Acta Crystallogr D Biol Crystallogr 52(Pt 6):1058–1066
9. Makino DL, Henschen-Edman AH, Larson SB, McPherson A (2007) Bence Jones KWR protein structures determined by X-ray crystallography. Acta Crystallogr D Biol Crystallogr 63:780–792

10. Baden EM, Owen BA, Peterson FC, Volkman BF, Ramirez-Alvarado M, Thompson JR (2008) Altered dimer interface decreases stability in an amyloidogenic protein. J Biol Chem 283(23):15853–15860

11. Kenniston JA, Faucette RR, Martik D, Comeau SR, Lindberg AP, Kopacz KJ, Conley GP, Chen J, Viswanathan M, Kastrapeli N, Cosic J, Mason S, DiLeo M, Abendroth J, Kuzmic P, Ladner RC, Edwards TE, TenHoor C, Adelman BA, Nixon AE, Sexton DJ (2014) Inhibition of plasma kallikrein by a highly specific active site blocking antibody. J Biol Chem 289(34):23596–23608

12. Kalinowska B, Banach M, Konieczny L, Roterman I (2015) Application of divergence entropy to characterize the structure of the hydrophobic core in DNA interacting proteins. Entropy 17(3):1477–1507

13. Roterman I, Banach M, Kalinowska B, Konieczny L (2016) Influence of the aqueous environment on protein structure—a plausible hypothesis concerning the mechanism of amyloidogenesis. Entropy 18(10):351

14. Levitt MA (1974) A simplified representation of protein conformations for rapid simulation of protein folding. J Mol Biol 104:59–107

15. Kullback S, Leibler RA (1951) On information and sufficiency. Ann Math Stat 22:79–86

16. Banach M, Prudhomme N, Carpentier M, Duprat E, Papandreou N, Kalinowska B, Chomilier J, Roterman I (2015) Contribution to the prediction of the fold code: application to immunoglobulin and flavodoxin cases. PLoS One 10(4):e0125098

17. Dygut J, Kalinowska B, Banach M, Piwowar M, Konieczny L, Roterman I (2016) Structural interface forms and their involvement in stabilization of multidomain proteins or protein complexes. Int J Mol Sci 17(10):E1741

18. Kalinowska B, Banach M, Konieczny L, Marchewka D, Roterman I (2014) Intrinsically disordered proteins – relation to general model expressing the active role of the water environment. Adv Protein Chem Struct Biol 94:315–346

19. Banach M, Kalinowska B, Konieczny L, Roterman I (2016) Sequence-to-structure relation in proteins-amyloidogenic proteins with chameleon sequences. J Proteomics Bioinform 9:264–275

20. Katzmann JA, Abraham RS, Dispenzieri A, Lust JA, Kyle RA (2005) Diagnostic performance of quantitative kappa and lambda free light chain assays in clinical practice. Clin Chem 51(5):878–881

21. Krol M, Roterman I, Drozd A, Konieczny L, Piekarska B, Rybarska J, Spolnik P, Stopa B (2006) The increased flexibility of CDR loops generated in antibodies by Congo red complexation favors antigen binding. J Biomol Struct Dyn 23(4):407–416

22. Schormann N, Murrell JR, Liepnieks JJ, Benson MD (1995) Tertiary structure of an amyloid immunoglobulin light chain protein: a proposed model for amyloid fibril formation. Proc Natl Acad Sci U S A 92(21):9490–9494

23. http://www.ks.uiuc.edu/Research/vmd/

24. Roterman I, Banach M, Konieczny L (2017) Application of fuzzy oil drop model describes amyloid as ribbonlike micelle. Entropy 19(4):167

25. Schütz AK, Vagt T, Huber M, Ovchinnikova OY, Cadalbert R, Wall J, Güntert P, Böckmann A, Glockshuber R, Meier BH (2015) Atomic-resolution three-dimensional structure of amyloid β fibrils bearing the Osaka mutation. Angew Chem Int Ed Engl 54(1):331–335

26. Król M, Roterman I, Piekarska B, Konieczny L, Rybarska J, Stopa B (2003) Local and long-range structural effects caused by the removal of the N-terminal polypeptide fragment from immunoglobulin L chain lambda. Biopolymers 69(2):189–200

27. Roterman I, Rybarska J, Monieczny L, Skowronek M, Stopa B, Piekarska B, Bakalarski G (1998) Congo red bound to α-1-proteinase inhibitor as a model of supramolecular ligand and protein complex. Computer Chem 22:61–70
28. Banach M, Konieczny L, Roterman I (2013) Can the structure of hydrophobic core determine the complexation site ? In: Roterman-Konieczna I (ed) Identification of ligand binding site and protein-protein interaction area. Springer Dordrecht, Heidelberg/New York/London, pp 41–54
29. BKChem (http://bkchem.zirael.org)
30. Serpell LC (2000) Alzheimer's amyloid fibrils: structure and assembly. Biochim Biophys Acta 1502(1):16–30
31. Lendel C, Bolognesi B, Wahlstrőm A, Dobson CM, Gräslund A (2010) Detergent-like interaction of Congo red with the amyloid β-peptide. Biochemistry 49:1358–1360

Chapter 6
Supramolecular Structures as Carrier Systems Enabling the Use of Metal Ions in Antibacterial Therapy

J. Natkaniec, Anna Jagusiak, Joanna Rybarska, Tomasz Gosiewski, Jolanta Kaszuba-Zwoińska, and Małgorzata Bulanda

Abstract The antimicrobial activity of metal ions, especially silver ions, has been known since ancient times. Consequently, finding an accessible, cheap and efficient carrier of metal ions remains an important challenge in molecular biology. The supramolecular system presented in this chapter consists of a mixture of Congo red and Titan yellow molecules forming a supramolecular ligand which is a potent complexing agent of silver ions. Delivery of ions in complex with supramolecular dye is advantageous due to the reduced toxicity. In addition, the use of Congo red provides selective action and – thanks to increased solubility – facilitates efficient dispersion of the carrier dye and excretion from the organism.

Keywords Supramolecular compounds • Congo red • Titan yellow • Silver ions • Microorganisms • Antimicrobial activity • Antifungal activity

The original version of this chapter was revised.
An erratum to this chapter can be found at https://doi.org/10.1007/978-3-319-65639-7_8

J. Natkaniec • T. Gosiewski • M. Bulanda
Department of Molecular Medical Microbiology, Chair of Microbiology,
Jagiellonian University – Medical College, Krakow, Poland
e-mail: joanna.natkaniec@uj.edu.pl; tomasz.gosiewski@uj.edu.pl; malgorzata.bulanda@uj.edu.pl

A. Jagusiak (✉) • J. Rybarska
Chair of Medical Biochemistry, Jagiellonian University – Medical College,
Kopernika 7, 31-034, Krakow, Poland
e-mail: anna.jagusiak@uj.edu.pl; mbstylin@cyf-kr.edu.pl

J. Kaszuba-Zwoińska
Department of Pathophysiology, Jagiellonian University – Medical College,
Czysta 18, 31-121, Krakow, Poland
e-mail: jolanta.kaszuba-zwoinska@uj.edu.pl

6.1 Introduction

The increasing antibiotic resistance of microorganisms is one of the greatest global health problems, as recalled by the World Health Organization [1, 2]. The percentage of multi-drug resistant isolates or isolates resistant to commonly used antibiotics is increasing throughout Europe. *Actinobacter baumani*, *Pseudomonas aeruginosa* and Enterobacteriaceae are included in the top three "critical" pathogens which have MDR (Multi-Drug Resistance). They are causing severe, frequently deadly infections and there is a critical need for new drugs to combat these pathogens [3–5]. In 2016, most strains – among Gram negative species Enterobacteriacae (such as: *Klebsiella pneumoniae* or *Escherichia coli*) that were also resistant to the third generation of cephalosporins, fluoroquinolones and aminoglycosides – have been reported in southern and southeastern Europe [6]. Methicillin-resistant *Staphylococcus aureus* (MRSA) is no longer just a hospital treatment problem, as in many parts of the world, including Europe, the number of MRSA infections (community-associated MRSA, CA-MRSA) is increasing. In Europe, eight out of 30 countries have reported having MRSA indexes exceeding 25% [6]. One of the causes of this problem is the high use of antibiotics, which increased significantly in the years 2010–2014 in the European Community, although it did not look the same in all countries. For example, in 2014, the number of antibiotics administered outside hospital treatments in Greece was more than three times higher than in the Netherlands (10.6 DDD (defined daily dose) per 1000 inhabitants per day in the Netherlands to 34.1 DDD per 1000 inhabitants per day in Greece) [7]. Unfortunately, the mere supervising of alarm strains, their registration, as well as the control system of antibiotic consumption are insufficient in the fight against microbial multi-resistance and searching for alternative means to solve the problem becomes necessary.

Among many propositions of solutions we will find e.g.: applying the so-called combined therapy, which uses two or more antibiotics administered simultaneously or combines the action of two different drugs in one molecule [8]. A promising alternative is the combination of chemotherapeutics with heavy metal ions, for example: silver, which has been known for its bactericidal properties since ancient times [9, 10]. The ideal situation would be to use silver compounds to increase the antibacterial effectiveness of commonly known and used antibiotics. However, in order for silver ions to be selective, they must be bound to a suitable carrier that would allow them to be precisely delivered and, in addition, would allow a more effective bacteriostatic/bactericidal effect when coacting with the antibiotic.

Finding an easily accessible, cheap and efficient carrier is one of the most important challenges researchers are facing [11, 12]. The carrier system presented here consists of a mixed supramolecular system of CR, which binds TY, which is a strong complexing agent for silver ions, by means of intercalation. The use of metal ions in the form of complexes makes it possible to lower the toxicity threshold. On the other hand, using a carrier system provides selectivity of action and, by increased solubility, facilitates excretion from the organism.

6.2 Silver Complexes with TY in the Supramolecular CR System

Supramolecularity is the phenomenon of the occurrence of organic compounds in the form of multimolecular systems resulting from the association of single molecules [13]. The classic example of a supramolecular system is CR (Fig. 6.1A). CR-type compounds associate creating elongated ribbon-like structures. Systems produced this way demonstrate the ability to bind different guest organic compounds by means of intercalation and, consequently, are characterized by various activities resulting from the nature of introduced components. Properly selecting the elements for such a complex system allows e.g. to increase its polarity and solubility, reduce cell entry and facilitate excretion, which further lowers the toxicity of components carried by the supramolecular system [14–16]. Supramolecular CR interacts with proteins, although it is an atypical ligand, which binds outside the active group of the protein [17–20]. There are also reports of bacteria interacting with CR [21–23].

Silver is an example of an antibacterial metal, commonly used locally, mainly as metallic silver (nanosilver, colloidal silver). Delivering silver as complexed ions using a carrier increases the safety of therapies using this metal. CR, however, has a low affinity for binding metals, including silver, hence the search for other compounds that form metal ion complexes with high efficiency. These compounds should interact directly with proteins, or indirectly through interaction with CR. One of the compounds which partially fulfils these parameters is TY (TY or DY9) (Fig. 6.1B) [24, 25]. TY poorly interacts with proteins compared to CR. It however creates strong complexes with CR, which can thus be used as a carrier system for both TY and its complex with silver ions. The TY molecule is symmetrical and planar, although its symmetry is disturbed by the central bond [26]. Similarly to CR, TY has a non-polar central benzene ring and polar sulphonic groups located at the poles of the molecule. This kind of structure allows for the self-association of individual molecules leading to the formation of complex supramolecular systems. A triazene bond is located at the centre of the TY molecule. This feature affects the

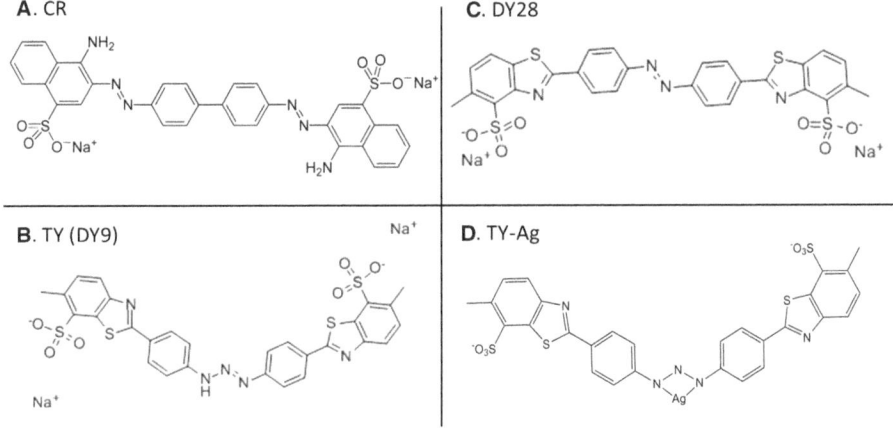

Fig. 6.1 Structural formulas: (**A**) CR; (**B**) TY; (**C**) DY 28; (**D**) TY–silver ions complex

spatial structure of TY, which is different than in the case of CR. The triazene moiety is involved in the formation of TY complexes with silver ions (Fig. 6.1D), as demonstrated by using the TY analogue – DY 28 which, instead of the triazole bond, has a diazo bond and does not form complexes with Ag⁺ ions (Fig. 6.1C).

The complexation of silver ions by TY was demonstrated by spectral analysis, and independently by using sodium dithionite to reduce ternary CR/TY/Ag⁺ complexes previously subjected to electrophoresis. As a result of reduction CR was decomposed and discoloured (because the azo bonds were reduced), the TY stain remained unchanged (as the triazene bond was not reduced), while silver appeared in the form of dark spots both within the TY/Ag⁺ complex and the CR/TY/Ag⁺ complex (both migrating towards the anode) as well as in the form of the excess migrating towards the cathode [24].

Silver ions are bound by TY in a molar ratio of 1:0.8, as demonstrated by titration of TY with silver (Fig. 6.2).

The strength of silver ions – TY interaction was evaluated by spectral analysis of the stability of the complexes in the presence of anions present in physiological solutions (Br^-, PO_4^-, Cl^-, CNS^-). Complexing silver ions increases the reactivity threshold of silver practically only for -SH groups, which are present on the surface of bacteria, because silver reactivity in complexes is reduced. The dissociation constant of the complex was estimated to be approx. 10^{-13}, because the anions with Kd values $<10^{-13}$ did not displace silver from the complex [25].

The viscosity of the TY/Ag⁺ complex was evaluated using viscometric analysis – the analysis of solution sedimentation in time, depending on composition. Viscosity

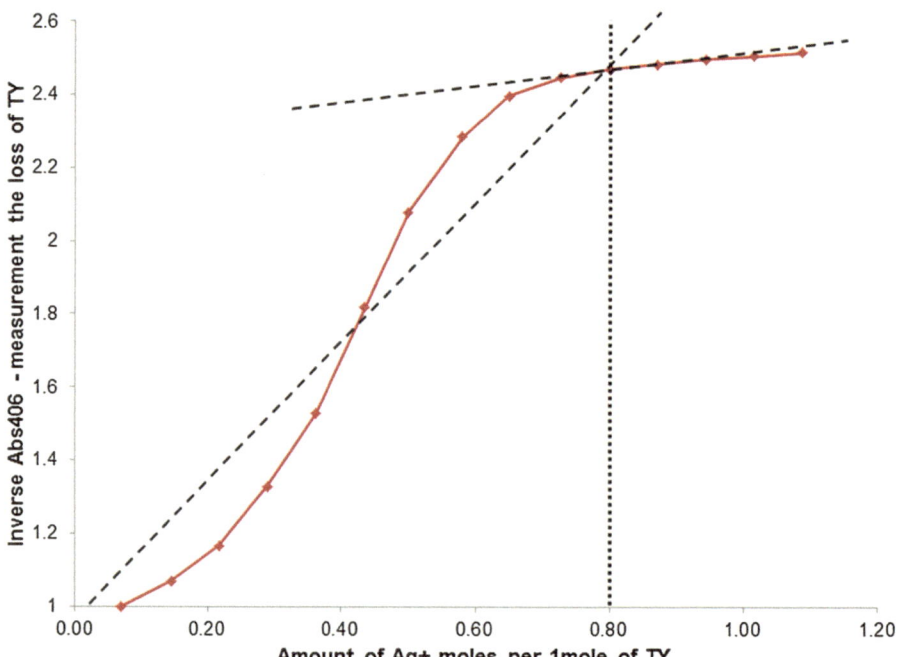

Fig. 6.2 Effect of titration of Titan yellow (TY) with silver (Ag⁺). The optimal molar ratio of TY:Ag⁺ is 1:0.8

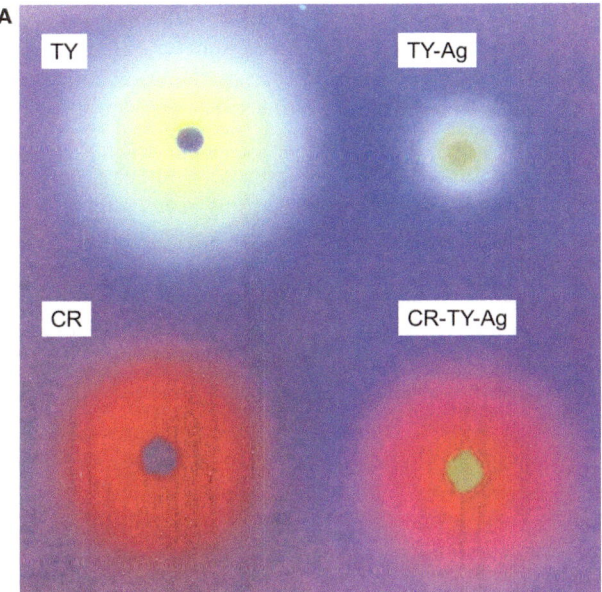

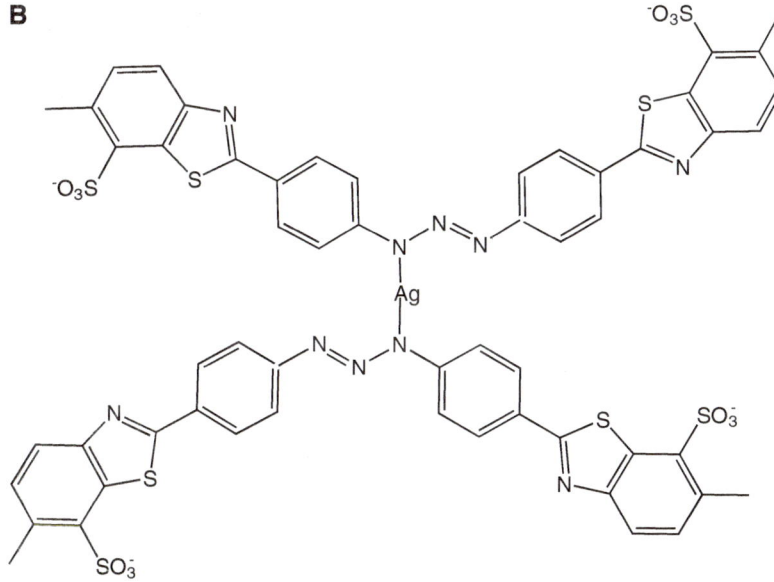

Fig. 6.3 (**A**) Comparison of diffusion rates in agarose gel: TY, CR, TY/Ag$^+$ and CR/TY/Ag$^+$ complexes after 24 h – a significant slowdown of the TY/Ag$^+$ complex diffusion due to increased viscosity of the complex results from the cross-linking of TY molecules by silver ion (**B**)

increases with the addition of an increasing amount of silver, hence the conclusion that silver cross-links TY molecules (Fig. 6.3B).

These results were confirmed in a diffusion test, which compared the diffusion of TY, CR as well as TY/Ag$^+$ and CR/TY/Ag$^+$ complexes in agarose gel over 24 h. The results have shown, that the TY/Ag$^+$ complex practically does not diffuse (diffusion zone diameter of 8 mm), and when bound to CR, the diameter of the diffusion zone is 18 mm and is close to the free TY (diffusion zone diameter of 20 mm) and free CR (diffusion zone diameter of 22 mm). This shows the important role of CR in increasing the solubility of the TY/Ag$^+$ complex (Fig. 6.3A).

Since the diffusion zone in the CR/TY/Ag$^+$ complex is 18 mm and is 4 mm smaller than the diffusion zone of free CR (which equals 22 mm), we conclude that the TY/Ag$^+$ complex is bound by CR. TY interacts with CR by intercalation into the ribbon structure of CR. The presence of CR in the CR/TY/Ag$^+$ complex reduces the viscosity of the complex, at the same time increasing its solubility. Additionally, the presence of CR may extend the range of TY activity by binding to certain bacteria [21–23]. Both TY and CR are carrying negative charges, yet they manage to interact and create stable complexes. This was confirmed in studies using electrophoretic and spectrophotometric methods [24, 25]. The results of electrophoresis on agarose gel show the formation of CR and TY complexes at an increasing CR to TY molar ratio. At two and five times molar excess of TY over CR (CR/TY4 and CR/TY5 samples) it was observed that the complex migrating quicker than free CR was formed and no excess of CR was observed. At 1:1 molar ratio and 2- and 5-fold excess of CR (CR/TY3, CR/TY2 and CR/TY1 samples), a smear of free CR was observed. Comparison of the migration speed of free TY with all the samples containing complexes (line A) indicates the presence of complexes in all five samples migrating at the same rate, as opposed to samples containing analogous amounts of free TY migrating at different rates (Fig. 6.4). The presence of Ag$^+$ complexed with TY increases the CR binding capability, because even at molar ratios of CR to the TY/Ag$^+$ complex equal to 5:1 and 2:1 (CR/TY/Ag1 and CR/TY/Ag2 samples) there is no free, unbound CR, as it was in the case of CR/TY complexes without silver (Fig. 6.5).

A distinct change in the absorption spectrum of TY after binding silver ions was observed (Fig. 6.6).

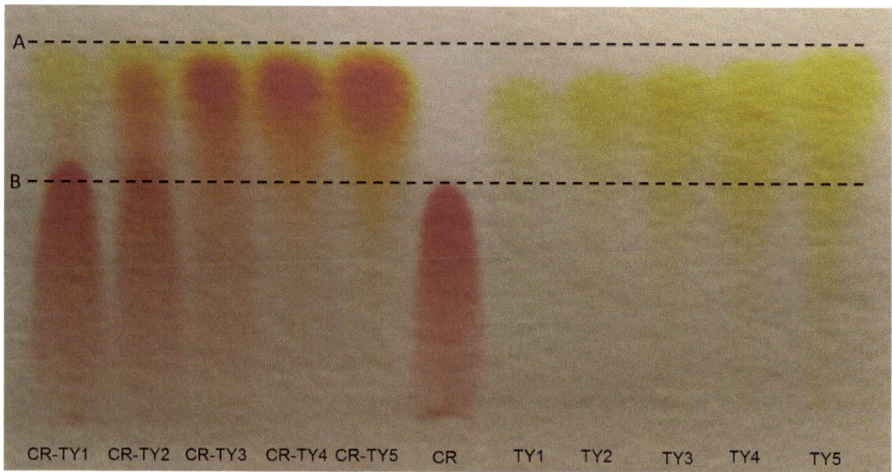

Fig. 6.4 Agarose gel electrophoresis of CR/TY molar ratio: CR/TY1 – 5:1; CR/TY2 – 2:1; CR/TY3 – 1:1; CR/TY4 – 1:2; CR/TY5 – 1:5; CR (equivalent to molar ratio 1:1); TY-Titan yellow; Molar ratio equivalents: TY1 – 5:1; TY2 – 2:1; TY3 – 1:1; TY4 – 1:2; TY5 – 1:5

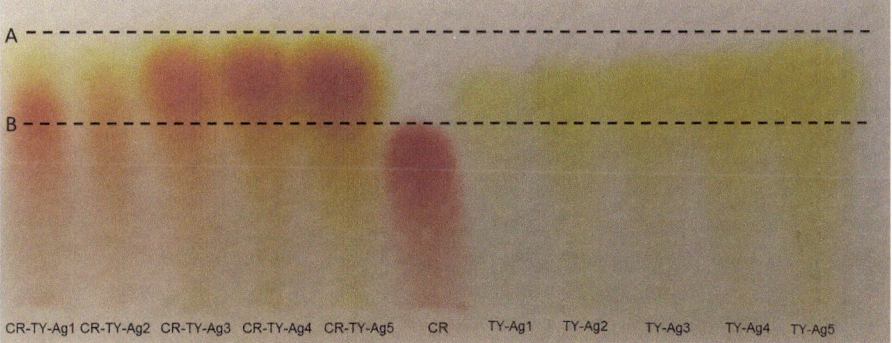

Fig. 6.5 Agarose gel electrophoresis of CR/TY/Ag⁺ molar ratio: CR/TY/Ag1 – 5:1; CR/TY/Ag2 – 2:1; CR/TY/Ag3 – 1:1; CR/TY/Ag4 – 1:2; CR/TY/Ag5 – 1:5; CR-Congo red (equivalent to molar ratio 1:1); TY/Ag⁺; Molar ratio equivalents: TY/Ag1 – 5:1; TY/Ag2 – 2:1; TY/Ag3 – 1:1; TY/Ag4 – 1:2

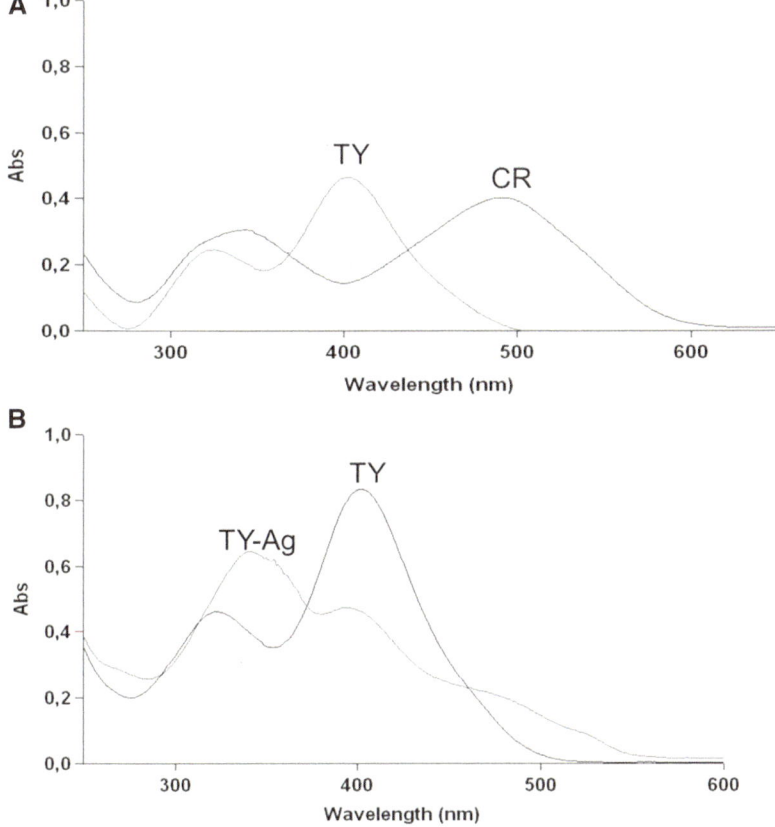

Fig. 6.6 Absorption spectra: (**A**) CR and TY; (**B**) TY and TY/Ag⁺

6.3 Antimicrobial and Antifungal Activity of Silver Ion Complexes with Titan Yellow (TY/Ag⁺) and TY/Ag⁺ Complexes with Congo Red (CR/TY/Ag⁺)

TY is a carrier system for silver ions. In the form of a complex it can affect the efficiency of antibiotics, by acting with them and, at the same time, reducing the toxicity of the used silver.

Free Ag^+ ions demonstrate high antimicrobial activity against Gram-positive and Gram-negative bacteria, fungi and certain viruses [27]. There are also theories regarding the mechanisms which inhibit the actions of silver on bacteria, such as: induction of oxygen free radicals or direct destruction of bacterial cell membranes by silver ions. Silver ions can interact with bacterial cell walls, bacterial capsule, cell membrane proteins, but also penetrate into the cytoplasm [28–30]. Silver ions block the activity of respiratory enzymes in bacterial cells, which prevents cellular respiration. Ag^+ ions have a high affinity to phosphate groups present in nucleic acids, thus inhibiting the replication of bacterial DNA [31–33]. Uncomplexed silver strongly binds to cell proteins through thiol, amino, carboxyl, imidazole and phosphate groups [34]. Silver, however, after complexation with the carrier, can practically bind only to the -SH groups. It is confirmed by the results presented in Table 6.1, which show a similar reactivity of silver-sensitive bacteria towards silver, as well as towards Ellman's reagent, which is specific for binding thiol groups. The amino acid found in the wall of bacteria susceptible to the activity of silver is cysteine, with a highly reactive -SH group [35, 36].

Excessive amount of silver introduced to the body accumulates in the tissues and can lead to necrosis of some organs, such as the liver. The concentration of silver ions over 10 mg/mL may be toxic for the organism [33], and the wide and uncontrolled use of silver can lead to developing bacterial resistance to silver, as currently is the case of antibiotics. It is therefore justified to search for silver carriers which allow its controlled release and also removal from the body [31].

In the nineteenth century the antiseptic properties of silver against microorganisms (*Staphylococcus aureus, Streptococcus, Pseudomonas; Escherichia*) were proven [37]. To date, silver nitrate has been shown to be effective against microorganisms of the genus *Neisseria* and *Pseudomonas*. The reduction of toxicity of silver used in therapy can be achieved by forming organic complexes. Among commonly known and used complexes are: silver sulfadiazine salt [38, 39] and silver complexes with sulfonamide drugs, which act against Gram-negative and Gram-positive bacteria [40]. Colloidal silver or immobilized silver nanoparticles e.g. on silica carriers, are also used. It seems however, that using supramolecular silver carriers is more effective because of the good solubility of the preparation, facilitated excretion and prevention of cell entry leading to the reduced toxicity. The proposed complex of TY with silver (TY/Ag⁺) is just this kind of an organometallic system with possible therapeutic potential and numerous advantages that indicate a chance of its application in medicine. One of the advantages is the use of ionic silver. The application of this complex limits the accumulation of the metal in the

Table 6.1 Screening of TY, CR, Ellman's Reagent (ER) and the TY/Ag$^+$ for selected strains – comparison with the antibiotic appropriate for the given strain: *AMC* amoxicillin with clavulanic acid (30 mg); *FEP* cefepime (30 mg); *NY* nystatin (100 IU)

Reference strains	NaCl	ANTIBIOTIC (zone)	TY (2.9 mM)	CR (2.9 mM)	ER (2.9 mM)	TY/Ag (Ag conc. 2.3mM)	TY/Ag (Ag conc. 4.6mM)
Staphylococcus epidermidis ATCC 700296	Ø	AMC (28)	Ø	Ø	Ø	11.4	12
Pseudomonas aeruginosa ATCC 27853	Ø	FEP (26)	Ø	Ø	Ø	10.7	11.5
Pseudomonas aeruginosa ATCC 33348	Ø	FEP (26)	Ø	Ø	Ø	9.3	10.7
Escherichia coli ATCC 25922	Ø	AMC (21)	Ø	Ø	Ø	10.5	11.3
Escherichia coli ATCC 35218	Ø	AMC (13)	Ø	Ø	14	9	9.5
Staphylococcus aureus ATCC 25923	Ø	AMC (35)	Ø	Ø	21	12	11
Staphylococcus aureus ATCC 29213	Ø	AMC (18)	Ø	Ø	Ø	8.7	10
Enterococcus faecalis ATCC 2912	Ø	AMC (26)	Ø	Ø.	Ø	Ø	Ø
Streptococcus agalactiae ATCC BAA-611	Ø	AMC (26)	Ø	Ø	Ø	Ø	Ø
Streptococcus pyogenes ATCC 700294	Ø	AMC (32)	Ø	Ø	Ø	Ø	Ø
Candida albicans ATCC 10231	Ø	NY (25)	Ø	Ø	Ø	14	16

Observation of the size of bacterial inhibition zones (mm) Gray indicates results where the zones were greater than zero body, due to the facilitation of excretion in the form of a complex. Complexation of silver reduces its toxicity and allows for its slow release, which prevents the formation of therapeutically disadvantageous silver agglomerates [33]. Interaction between TY and CR creates a possibility of interaction of the ternary TY/CR/Ag$^+$ system with certain proteins, and also leads to the increased solubility of the complex. Another advantage is the specificity expressed by the reactivity of silver ions towards thiol groups.

6.3.1 Methodology

In order to evaluate the effect of silver complexed with TY or with the CR/TY system we have performed screening tests on gram positive and gram negative bacteria and on yeast-like fungi.

The complexes were tested using strains from the collection of the Department of Microbiology of the Jagiellonian University Medical College. In the case of *Staphylococcus aureus*, *Escherichia coli* and *Pseudomonas aeruginosa*, reference strains recommended by EUCAST (European Committee on Antimicrobial Susceptibility Testing) were chosen as control strains for susceptibility testing and as control strains in quality control. These strains are: *Staphylococcus aureussubsp. aureus* (ATCC® 29213™), *Escherichia coli* (ATCC® 25922™) – a susceptible strain that does not possess any antibioticresistance, *Escherichia coli* (ATCC® 35218™) – a strain producing TEM-1β-lactamase, which causes resistance to all β-lactam antibiotics, *Pseudomonas aeruginosa* (ATCC® 27853™). Furthermore one strain of *Staphylococcus aureussubsp. aureus* (ATCC® 25923™), which is recommended by CLSI (Clinical & Laboratory Standards Institute) for susceptibility testing, was chosen.Other reference strains: *Staphylococcus epidermidis* (ATCC® 700296™), *Pseudomonas aeruginosa* (ATCC® 33348™), *Enterococcus faecalis* (ATCC® 2912™), *Streptococcus agalactiae* (ATCC® BAA-611™), *Streptococcus pyogenes* (ATCC® 700294™), *Candida albicans* (ATCC® 10231™) were selected according to their availability, as were the clinical strains: *Pseudomonas aeruginosa* (1815), *Pseudomonas aeruginosa* (18168), *Pseudomonas aeruginosa* (22726), *Escherichia coli* (E1), *Escherichia coli* (E123), *Escherichia coli* (M243), *Staphylococcus aureus* (277), *Staphylococcus aureus* (1934), *Staphylococcus aureus* (26265).

A diffusion-disk screening method was used to test for antibacterial and antifungal activity by using a filter paper disc impregnated with the tested compound at a specific concentration. It has been assumed that the tested compound will behave similarly to the antibiotic-impregnated disc in the Kirby-Bauer diffusion-disc method, which is used for the determination of drug-susceptibility of microbes. The diffusion process is radial, so a concentration gradient is created. The highest concentration of the tested antibiotic is at the edge of the disc and decreases with the distance. The diameter of the microbial growth inhibition zone is directly proportional to the antibiotic susceptibility, i.e. the greater the growth inhibition zone around the disc, the more the microorganism is susceptible to the given antibiotic.

Depending on the size of this zone, microbes are defined as: sensitive, medium sensitive or resistant, based on the accepted standards (EUCAST recommendations).

Ten microlitres of the tested compound were applied onto sterile filter paper discs, and then placed on MHA (Mueller-Hinton Agar) media inoculated with bacterial suspension or Sabouraud medium inoculated with fungi suspension (optical density of the suspensions was 0.5 McFarland). The compound diffused into the medium and, if a microbial growth inhibition zone was observed around the disc with a diameter at least 1 mm greater than the diameter of the disc impregnated with the test compound, it was treated as a positive effect of the tested compound.

First, the control system was tested using free TY without the complexed metal and free CR. On the media, inoculated with the suspension of bacteria or fungi strains prepared in a standard way, were placed two sterile filter paper discs (Ø, diameter = 5 mm) impregnated with 0.9% NaCl (control), the tested compound, i.e.: TY (2.9 mM) or CR (2.9 mM) and, additionally, a disc with an antibiotic, selected so to have the widest spectrum of antimicrobial activity possible. As for the strains: amoxicillin with clavulanic acid (AMC) was used at the concentration of 30 mg *for* *S. aureus, E. coli, Enterococcus faecalis,Streptococcus pyogenes,Streptococcus agalactiae*; a disc with cefepime at the concentration of 30 mg was used for *Pseudomonas aeruginosa*; Nystatin was used at the concentration of 100 IU for *Candida albicans*.

Subsequently, two concentrations of the TY/Ag^+ complex were tested (concentrations of TY: 2.9 mM and 5.75 mM; Ag^+ concentrations respectively: 2.3 mM and 4.6 mM; molar ratio of $TY:Ag^+$ = 1:0.8). The $CR/TY/Ag^+$ ternary complex was also tested (two concentrations of TY and CR: I. TY: 2.9 mM; CR: 2.9 mM, Ag^+: 2.3 mM; II. TY: 5.75 mM; CR: 5.75 mM, Ag^+: 4.6 mM; in both cases the molar ratio of $TY:Ag^+$ = 1:0.8; the molar ratio of CR:TY 1:1). Sterile paper discs (Ø = 5 mm) were used, impregnated respectively with a filter solution of TY silver complex (TY/Ag^+), in two tested concentrations or in a solution of $CR/TY/Ag^+$. A disc impregnated with 0.9% NaCl was used as control. The discs thus prepared were placed on the MHA medium, inoculated with a bacterial suspension and on a Sabouraud medium, inoculated with a fungi suspension. Microbial suspensions were prepared from reference (Table 6.1) and clinical (Table 6.2) strains and each had an optical density of 0.5 McFarland.

The selective activity of Ag^+ complexed with TY or with CR/TY was assessed by tests, which consisted of comparing the effects of TY/Ag^+ and $CR/TY/Ag^+$ and the effects of the Ellman reagent (ER). Ellman's reagent is 5,5'-dithiobis (2-nitrobenzoic acid), abbreviated as DTNB – which binds and blocks the thiol groups. The concentration of ER used was 2.9 mM, which corresponds to the molar concentration of silver used. The procedure of preparing the microbial suspensions was performed as before. On the inoculated plates with the MHA or Sabourand medium, a control disc impregnated in 0.9% NaCl, a disc with one of the four aforementioned compounds and a disc with the antibiotic were applied.

Table 6.2 Results of the screening test of TY, CR, Ellman's Reagent (ER) and TY/Ag⁺ for selected clinical strains. Shaded cells indicate results with zones above zero

Clinical strains	NaCl	ANTIBIOTIC (zone)	TY (2.9 mM)	CR (2.9 mM)	ER (2.9 mM)	TY/Ag (Ag conc. 2.3mM)	TY/Ag (Ag conc. 5.7mM)
P.aeruginosa 1815	Ø	FEP (25)	Ø	Ø	Ø	8	9
P. aeruginosa 18168	Ø	FEP (15)	Ø	Ø	Ø	8	8
P. aeruginosa 22726	Ø	FEP (14)	Ø	Ø	Ø	8	10
E.coli E1	Ø	AMC (24)	Ø	Ø	13	9	8
E.coli 123	Ø	AMC (9)	Ø	Ø	8	6	11
E.coli M243	Ø	AMC (21)	Ø	Ø	12	6	6
S.aureus 277	Ø	AMC (28)	Ø	Ø	14	6	10
S.aureus 1934	Ø	AMC (21)	Ø	Ø	14	10	10
S.aureus 26265	Ø	AMC (20)	Ø	Ø	14	10	9

6.3.2   Results

In both cases, where free compounds were used: TY and CR did not show growth inhibition zones either in the test strains tested (Table 6.1) or in clinical strains (Table 6.2).

Growth inhibition zones after the application of different concentrations of Ag⁺ complexed with TY were observed in the case of the strains: *S. aureus, E. coli, S. epidermidis, P. aeruginosa, C. albicans*. No growth inhibition zones were observed for the strains: *E. faecalis, S. agalactiae, S. pyogenes* (Tables 6.1 and 6.2). Similar results were obtained for CR/TY/Ag⁺ complexes.

Growth inhibitotion zones after applying ER and TY/Ag⁺ as well as CR/TY/Ag⁺ were observed for the following reference strains: *E. coli* ATCC 35218 and *S. aureus* ATCC 25923 (Table 6.1) and in the case of the following clinical strains: *E. coli* E1; *E. coli* 123; *E. coli* 243, *S. aureus* 277; *S. aureus* 1934 and *S. aureus* 26265. Based on these results it can be assumed that complexed silver inhibits these bacterial strains by blocking the thiol groups.

No growth inhibition zones were observed after applying ER, TY/Ag⁺ and CR/TY/Ag¹ in the case of the following standard strains: *E. faecalis* ATCC 2912, *S. agalactiae* BAA-611, *S. pyogenes* ATCC 700294 and in the case of clinical strains: *E. fecalis* E1; E2; *S. agalactie* 135, 136, 149; *S. pyogenes* G0, 282, 287.

In the case of reference strains *S. epidermidis* ATCC 700296, *P. aeruginosa* ATCC 27853 and ATCC 33348, *E. coli* ATCC 25922, *S. aureus* ATCC 29213 and *C. albicans* ATCC 10231, as well as in the case of the clinical strains: *P. aeruginosa* 1815, 18168, 22726, and for the higher concentrations of TY/Ag⁺ for *C. tropicalis, C. parapsilosis* i *C. lusitaniae* growth inhibition zones were observed after using complexed Ag⁺, whereas such zones were not observed after applying ER. Probably, in the case of these strains, silver affects groups other than thiol groups (e.g. amino groups).

In experiments using the CR/TY/Ag⁺ complex, there was a tendency to increase the bacterial growth inhibition zone as compared to the experiments, where Ag⁺ was complexed only with TY. This tendency is evident especially when lower concentration of Ag (2.3 mM) in the CR/TY complex is used. CR significantly reduces the viscosity of the TY/Ag⁺ complex and improves its diffusion. This result, especially in conjunction with the tendency of CR to interact with certain bacteria, is the basis for recognizing the presented CR/TY system to be an effective silver ion carrier.

6.4   Determination of the Effect of TY/Ag⁺ and CR/TY/Ag⁺ on Antibiotic Activity on Selected Standard and Clinical Bacterial Strains

The increasing resistance of bacteria to standard antibiotics limits the possible use of this class of drugs and significantly narrows the therapeutic options. Hence, further studies have tested the possibility of using silver in new, yet unresearched TY/

Table 6.3 Observation of the size of growth inhibitory zones (mm)

Reference strains	Antibiotic	(1)	(2)	(3)	(4)
*P. aeruginosa*ATCC 27853	FEP	Ø	26	11	27
S. aureus ATCC 29213	AMC	Ø	18	10	28
E. coli ATCC 35218	AMC	Ø	13	11	22

(1) Buffer control (sterile discs with a 0.05 M TRIS-HNO$_3$ solution, pH 8.2), (2) sterile discs with antibiotics (3) sterile discs with the TY/Ag$^+$ solution, and (4) sterile discs with antibiotic with the solution (TY/Ag$^+$)

Antibiotics used: *AMC* amoxicillin with clavulanic acid (30 mg); *FEP* cefepim (30 mg)

Ag$^+$ and CR/TY/Ag$^+$ complexes as a possible support for antibiotics (an adjuvant system in antibiotic therapy). The effect of analysed silver complexes on antibiotics used for selected strains was tested.

Among the strains that gave a positive response of the growth inhibition zone after applying the TY/Ag$^+$ complex, three strains of standard reference microorganisms were selected for antibody testing: *Pseudomonas aeruginosa* ATCC 27853, *Staphylococcus aureus* ATCC 29213 *andEscherichia coli* ATCC 35218. The study was conducted in accordance with the methodology described above. Sterile filter paper discs (without or with antibiotics) were impregnated using a selected solution: (1) disc without antibiotic impregnated with a 0.05 M TRIS-HNO$_3$ solution, pH 8.2 (buffer control), (2) antibiotic disc, (3) disc without the antibiotic with a solution of the tested compound, and (4) antibiotic disc impregnated with the solution of test compound. The test compound used was: TY- silver ions complex (TY/Ag$^+$). Then the discs were applied to plates with MHA media inoculated with bacterial suspension.

The studies showed that compounds combined with silver ions (TY/Ag$^+$) create a growth inhibition zone. It was concluded that silver ions complexed with in TY does not inhibit the action of any of the tested antibiotics used for the analysed bacterial strains. Moreover, when comparing the growth inhibition zones of *E. coli* and *S. aureus* strains in the presence of antibiotic and the antibiotic combined with the test compound (TY/Ag$^+$), an increase of inhibitory action was observed for the combination. This result shows that silver in the presented complex acts as an adjuvant on the antibiotic used (Table 6.3 and Fig. 6.7).

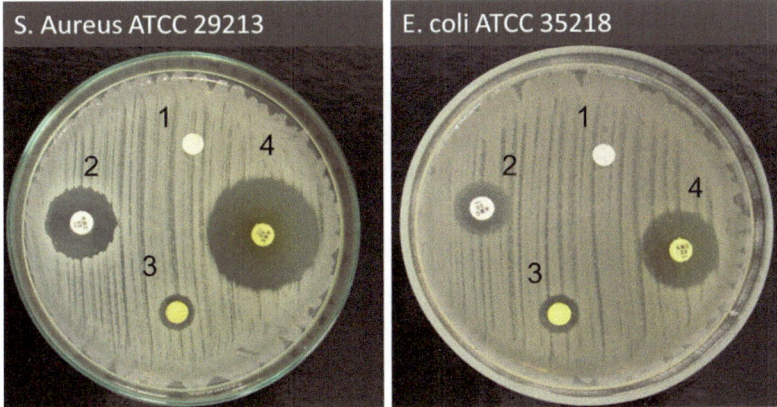

Fig. 6.7 Diameter of growth inhibition zones of *S. aureus* ATCC 29213 and *E. coli* ATCC 35218; (*1*) buffer control (sterile discs impregnated with a sterile 0.05 M TRIS-HNO3 solution, pH 8.2), (*2*) sterile discs with antibiotic (*3*) sterile discs impregnated with an TY/Ag⁺ solution and (*4*) sterile discs with antibiotic impregnated with the TY/Ag⁺ solution

6.5 Cytotoxicity Analysis of the TY/Ag⁺ Complex

Induction of the apoptosis process is one of the main goals of the designed cancer therapies. This is due to the developed mechanisms of cancer cell resistance to programmed death. Therefore, in the next stage of study, the cytotoxicity of the compounds used was analysed by comparing early apoptosis, late apoptosis and necrosis of U937 cells after application of the following concentrations of the TY/Ag⁺ complexes: 1 μg/mL; 50 μg/mL; 200 μg/mL and 400 μg/mL in a 96 h culture. It was shown that 1 μg/mL and 50 μg/mL concentrations of TY do not lead to cytotoxic effects in the studied cell line. The concentration of 50 μg/mL of TY/Ag⁺ results in late apoptosis while concentrations of 200 and 400 μg/mL cause necrosis.

Human U937 lymphoid cell line was obtained from the American Cell Culture Collection (ATCC, Rockville, MD) and cultured in RPMI 1640 medium (Gibco-BRL, USA), supplemented with 10%(v/v) fetal calf serum (Gibco-BRL, USA) heat inactivated, L-glutamine 0.2 M and gentamycin 50 μg/mL (Sigma-Aldrich, Germany) at 37 °C in a 5% CO_2 incubator of 90% humidity. Cell viability was monitored by trypan blue exclusion method and counted with a haemocytometer (Fuchs-Rosenthal chamber). The experiments were performed on cells in the logarithmic phase of growth under the condition of 98% viability, as assessed by trypan blue exclusion. U937 cells were passaged every 4 days. For the experiments U937 cells were seeded into 96-well (Nunck, Denmark) culture plates and grown at density of 0.5×10^6 cells/mL in threefold repetition.

24 h after cell passage of U937 the cell line cultured at density 0.5×106 cells/mL, TY, TY/Ag⁺ or ER reagents suspended in cell culture medium were added to the cell cultures.

The final concentration for TY, TY/Ag⁺ and ER was 50 μg/ml. Cells without any reagent treatment constituted a control group.

After 72 h of cell culture duration with the presence of added reagents, U937 cells were harvested by centrifugation at 280 g for 10 min and used for flow cytometry analysis.

After 96 h lasting U937 cell cultures, cells were harvested, washed twice with cold PBS (Sigma-Aldrich, Germany) and resuspended in 1× binding buffer (BD, Pharmingen TM, USA) at a concentration of 1× 10^6 cells/mL. Then, the solution (100 µL) was transferred to a 5-mL culture tube, and AnV-APC(AnV-APC, BD, Pharmingen TM, USA) and PI (PI, BD, Pharmingen TM, USA) were added, 5 µL each. The cells were vortexed gently and incubated in darkness at room temperature for 15 min. After adding 1× binding buffer (400 µL), the cells were analyzed on a FACS CALIBUR flow cytometer (Becton Dickinson, San Jose, CA) using Cell-Quest software to calculate the proportion of ADSCs representing various types. Unstained cells, cells stained with AnV-APC alone (for FL-4 fluorescence) and cells stained with PI alone (detected in FL-3) were used as controls to set up compensation and appropriate quadrants. At least 10,000 events were acquired for each sample.

The percentage of apoptotic cells was determined with allophycocyanin (APC) conjugated annexin V. Propidium iodide was used as a standard flow cytometric viability probe to distinguish necrotic cells from viable ones. AnV-APC-positive, PI-negative cells (AnV+PI-) were classified as early apoptotic, AnV-APC- and PI-positive cells (AnV+PI+) as late apoptotic, and AnV-APC-negative, PI-positive cells (AnV-PI+) as necrotic.

Under the experimental conditions, no early apoptotic cells were observed 72 h after the addition of reagents (Fig. 6.8). No late apoptosis and necrosis were observed in the control cells and the ones treated with the Ellman reagent or free TY. Cells treated with TY/Ag⁺ complex showed late apoptotic effects (Fig. 6.9) with a small percentage of necrotic cells (Fig. 6.10). The ability to induce apoptosis in tumor cells by the TY/Ag⁺ complex allows to consider it a promising potential anti-tumor system.

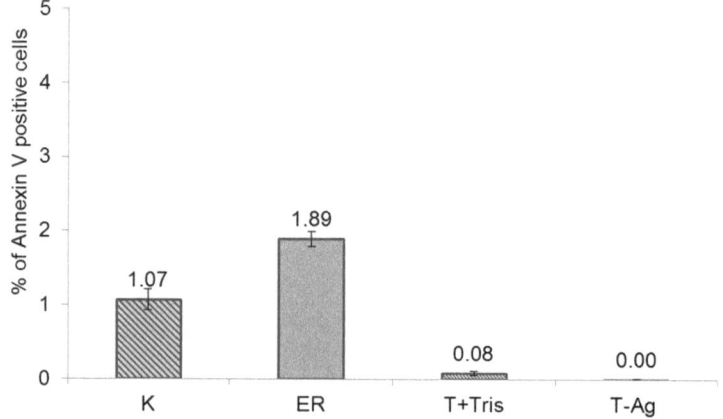

Fig. 6.8. Early apoptosis of U937 cells cultivated in vitro for 72 h with three different reagents (K - control, ER-Ellmans Reagent 50 µg/mL; T + Tris – Titan yellow 50 µg/mL and T-Ag⁺ – Titan yellow (50 µg/mL) – silver ions complex1:0.8 molar ratio). Data expressed as mean (±SD)

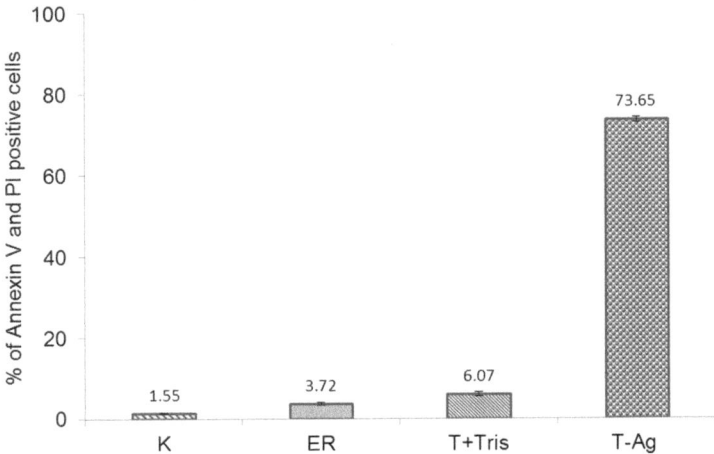

Fig. 6.9 Late apoptosis and necrosis of U937 cells cultivated in vitro for 72 h with three different reagents (K – control; ER-Ellmans Reagent; T + Tris – Titan yellow and T-Ag$^+$ – Titan yellow-silver ions complex). Data is expressed as mean (±SD)

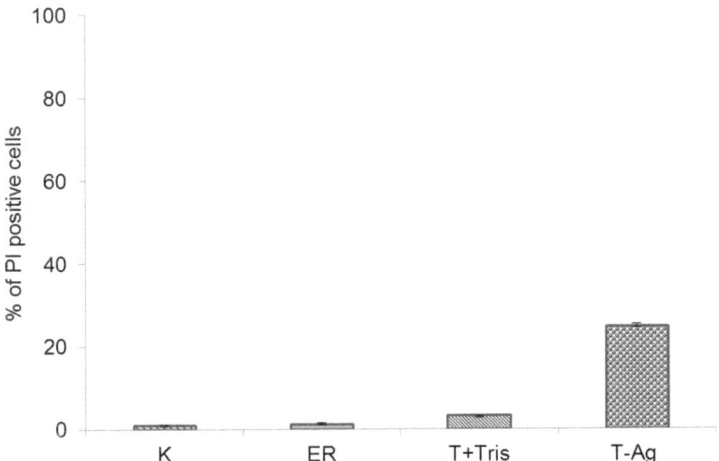

Fig. 6.10 Necrosis of U937 cells cultivated in vitro for 72 h with three different reagents (K – control; ER-Ellmans Reagent; T + Tris – Titan yellow and T-Ag$^+$ – Titan yellow-silver ions complex); Data is expressed as mean (±SD)

6.6 Summary

The complexes TY/Ag$^+$ and the ternary complex CR/TY/Ag$^+$ are systems with potentially therapeutic effects. The illustrated carrier systems, thanks to their very good solubility, reduce the accumulation of metal in the body by facilitating the elimination of silver in the form of a complex. Silver complexation reduces toxicity and allows to target its action mainly at interactions with thiol groups. Due to the interaction between TY and CR, the TY/CR/Ag⁺ system can additionally interact with different bacteria. Supramolecular CR protects against cell entry and reduce toxicity of silver ions, while increasing the selectivity of action. Thanks to these features the presented system can be more easily excreted from the body. No toxicity of free TY has been found, however it can induce apoptosis following the complexation of silver ions. Studies conducted on bacterial and fungal strains show that the complex exhibits antimicrobial activity. This is especially valuable in case of resistant bacterial strains that exhibit a significant degree of insensitivity to commonly used antibiotics; using the tested complex increases the extent of action of these antibiotics, consequently allowing them to be used again in the treatment of resistant microorganisms. The obtained results provide the basis for further research on the use of the presented organometallic complex in medicine as an antibacterial agent.

Acknowledgements We acknowledge the financial support from the National Science Centre, Poland (grant no. 2016/21/D/NZ1/02763) and from the project Interdisciplinary PhD Studies "Molecular sciences for medicine" (co-financed by the European Social Fund within the Human Capital Operational Programme) and Ministry of Science and Higher Education (grant no. K/DSC/001370).

References

1. WHO (2015) Antibiotic Resistance: Multi-country public awareness survey. ISBN 978 92 4 150981 7
2. WHO (2014) Antimicrobial resistance. global Report on Surveillance. ISBN 978 92 4 156474 8
3. News at a glance Science, 355(6328), 2017, 890–892
4. Taconelli E, Margini N (2017) WHO "Global priority list of antibiotic-resistant bacteria to guide research, discovery, and development of new antibiotics
5. Willyard C (2017) Drug-resistant bacteria ranked. Nature 543(7643):15
6. European Centre for Disease Prevention and Control (2016) Summary of the latest data on antibiotic resistance in the European Union Stockholm: ECDC
7. European Centre for Disease Prevention and Control (2015) Summary of the latest data on antibiotic consumption in the EU Antibiotic consumption in Europe Stockholm: ECDC
8. Wang KK, Stone LK, Lieberman TD et al (2016) A hybrid drug limits resistance by evading the action of the multiple antibiotic resistance pathway. Mol Biol Evol 33(2):492–500
9. Silver S, Phung LT, Silver G (2006) Silver as biocides in burn and wound dressings and bacterial resistance to silver compounds. J Ind Microbiol Biotechnol 33(7):627–634
10. Durán N, Marcato PD, De Conti R et al (2010) Potential use of silver nanoparticles on pathogenic bacteria, their toxicity and possible mechanisms of action. J Braz Chem Soc 21(6):949–959
11. Yang G-W, Gao G-Y, Wang C et al (2008) Controllable deposition of Ag nanoparticles on carbon nanotubes as a catalyst for hydrazine oxidation. Carbon N Y 46(5):747–752
12. Guo X, Mei N (2014) Assessment of the toxic potential of graphene family nanomaterials. J Food Drug Anal 22(1):105–115

13. Skowronek M, Stopa B, Konieczny L et al (1998) Self-assembly of Congo red – a theoretical and experimental approach to identify its supramolecular organization in water and salt solutions. Biopolymers 46(5):267–281
14. Rybarska J, Piekarska B, Stopa B et al (2001) Evidence that supramolecular Congo red is the sole ligation form of this dye for L chain lambda derived amyloid proteins. Folia Histochem Cytobiol 39(4):307–314
15. Zemanek G, Rybarska J, Stopa B et al (2003) Protein distorsion-derived mechanism of signal discrimination in monocytes revealed using Congo red to stain activated cells. Folia HistochemCytobiol 41(3):113–124
16. Stopa B, Piekarska B, Jagusiak A et al (2011) Acta Biochim Pol 58(Suppl. 2):282. Supramolecular Congo red as a potential drug carrier. Properties of Congo red-doxorubicin complexes in Proceedings of the 2nd Congress of Biochemistry and Cell Biology 46th Meeting of the Polish Biochemical Society and 11st Conference of the Polish Cell Biology Society, Kraków, 2011
17. Piekarska B, Drozd A, Konieczny L et al (2006) The indirect generation of long-distance structural changes in antibodies upon their binding to antigen. Chem Biol Drug Des 68(5):276–283
18. Rybarska J, Piekarska B, Stopa B et al (1984) In vivo accumulation of self-assembling dye Congo red in an area marked by specific immune complexes: possible relevance to chemotherapy. Folia HistochemCytobiol 42(2):101–110
19. Spólnik P, Stopa B, Piekarska B et al (2007) The use of Rigid, Fibrillar Congo red nanostructures for scaffolding protein assemblies and inducing the formation of amyloid-like arrangement of molecules. Chem Biol Drug Des 70(6):491–501
20. Roterman I, Rybarska J, Konieczny L et al (1998) Congo red bound to α-1-proteinase inhibitor as a model of supramolecular ligand and protein complex. Comput Chem 22(1):61–70
21. Arnold JW, Shimkets LJ (1988) Inhibition of cell-cell interactions in Myxococcus xanthus by congo red. J Bacteriol 170(12):5765–5770
22. Qadri F, Hossain SA, Ciznár I et al (1988) Congo red binding and salt aggregation as indicators of virulence in Shigella species. J. Clin Microbiol 26(7):1343–1348
23. Ishiguro EE, Ainsworth T, Trust TJ et al (1985) Congo red agar, a differential medium for Aeromonas salmonicida, detects the presence of the cell surface protein array involved in virulence. J. Bacteriol 164(3):1233–1237
24. Chłopaś K, Jagusiak A, Konieczny L et al (2015) The use of Titan yellow dye as a metal ion binding marker for studies on the formation of specific complexes by supramolecular Congo red. Bio-Algorithms Med-Syst 11(1):9–17
25. Rybarska J, Konieczny L, Jagusiak A et al (2017) Silver ions as EM marker of congo red ligation sites in amyloids and amyloid-like aggregates. Acta Biochim Pol 64:161–169
26. Stopa B, Piekarska B, Konieczny L et al (2003) The structure and protein binding of amyloid-specific dye reagents. Acta Biochim Pol 50(4):1213–1227
27. Zhao G, Stevens SE Jr (1998) Multiple parameters for the comprehensive evaluation of the susceptibility of Escherichia coli to the silver ion. Biometals 11(1):27–32
28. Lemire JA, Kalan L, Bradu A et al (2015) Silver oxynitrate, an unexplored silver compound with antimicrobial and antibiofilm activity. Antimicrob Agents Chemother 59(7):4031–4039
29. Hajipour MJ, Fromm KM, Akbar A et al (2012) Antibacterial properties of nanoparticles. Trends Biotechnol 30(10):499–511
30. Kim JS, Kuk E, Yu KN et al (2007) Antimicrobial effects of silver nanoparticles Nanomedicine Nanotechnology. Biol Med 3(1):95–101
31. Gupta A, Phung LT, Taylor DE et al (2017) Diversity of silver resistance genes in IncH incompatibility group plasmids. Microbiology 2046(147):42–3393
32. Matsumura Y, Yoshikata K, Kunisaki S et al (2003) Mode of bactericidal action of silver zeolite and its comparison with that of silver nitrate. Appl. Environ Microbiol 69(7):4278–4281
33. Schierholz JM, Lucas LJ, Rump A, Pulverer G (1998) Efficacy of silver-coated medical devices. J Hosp Infect 40(4):257–262

34. Russell AD, Hugo WB (1994) Antimicrobial Activity and Action of Silver. Prog Med Chem 31:351–370
35. Liau SY, Read DC, Pugh WJ et al (1997) Interaction of silver nitrate with readily identifiable groups: relationship to the antibacterialaction of silver ions Lett. Appl Microbiol 25(4):279–283
36. Feng QL, Wu J, Chen GQ et al (2000) A mechanistic study of the antibacterial effect of silver ions on Escherichia coli and Staphylococcus aureus. J Biomed Mater Res 52:662–668
37. Klasen HJ, Bauer K-H, Gravens DL et al (2000) Historical review of the use of silver in the treatment of burns. I Early uses. Burns 26(2):117–130
38. Silver S (2003) Bacterial silver resistance: molecular biology and uses and misuses of silver compounds. FEMS Microbiol. Rev 27(2–3):341–353
39. Percival SL, Bowler PG, Russell D et al (2005) Bacterial resistance to silver in wound care. J Hosp Infect 60(1):1–7
40. Ruutu M, Alfthan O, Talja M et al (1985) Cytotoxicity of latex urinary catheters. Br J Urol 57(1):82–87

Chapter 7
Congo Red Interactions with Single-Walled Carbon Nanotubes

Anna Jagusiak, Barbara Piekarska, Katarzyna Chłopaś, and Elżbieta Bielańska

Abstract A new method of dispersion of single-wall carbon nanotubes (SWNT) in aqueous solution using supramolecular compounds is proposed in this chapter. The described system consists of SWNT overlaid by Congo red. SWNT are formed from a rolled layer of graphene, providing a large surface area for binding compounds with planar, aromatic structures (including drugs). Congo red is able to associate with proteins in the form of supramolecular, ribbon-like structures, and may bind various drugs by intercalation.

The study reveals strong interactions between Congo red and the surface of SWNT. The authors' aim was to explain the mechanism driving this interaction. Spectral analysis of the SWNT-CR complex, effects of sonication on CR binding, microscopic imaging and molecular modelling analyses are all discussed. Results indicate that binding of supramolecular Congo red to the surface of nanotubes is based on face-to-face stacking. Having attached itself to the surface of a nanotube, a dye molecule may attract other similarly oriented molecules, giving rise to a protruding supramolecular appendage. This explains the high affinity of CR for nanotubes and the resulting system's capability to bind drugs.

Analysis of complexes formed by SWNT-CR with the model drug (DOX) and with other planar compounds (EB, TY) indicates that it may be possible to construct complexes capable of binding multiple compounds simultaneously.

Keywords Single-walled carbon nanotubes • Supramolecular compounds • Bis-azo dye • Congo red • Chemical container • Drug delivery system • Shortenng of carbon nanotubes • Pi-pi stacking • Face-to-face stacking

A. Jagusiak (✉) • B. Piekarska • K. Chłopaś
Chair of Medical Biochemistry, Jagiellonian University – Medical College,
Kopernika 7, 31-034, Krakow, Poland
e-mail: anna.jagusiak@uj.edu.pl; mbpiekar@cyf-kr.edu.pl; katarzyna.chlopas@wp.pl

E. Bielańska
Institute of Catalysis and Surface Chemistry, Polish Academy of Science,
Niezapominajek 8, 30-239, Krakow, Poland
e-mail: ncbielan@cyf-kr.edu.pl

© The Author(s) 2018
I. Roterman, L. Konieczny (eds.), *Self-Assembled Molecules – New Kind of Protein Ligands*, https://doi.org/10.1007/978-3-319-65639-7_7

7.1 Methods of Dispersing Carbon Nanotubes – The Search for Optimal Dispersion Methods

Carbon nanotubes (CNT) are used in electronics [1], composite materials research [2, 3], catalysis [4], textiles [5] and many other areas. A particularly interesting is their use in the field of biomedicine: in biosensors [6, 7], medical diagnostics [8, 9], transplantology [10, 11], tissue engineering [12] or in pharmacology – including the field of targeted therapies [13–16].

In studies on the use of CNT in biomedicine it is necessary to obtain their suspension in water media. CNT are most often found not in the form of single fibres, but in the form of bundles stabilized by van der Waals interactions and it is very difficult to evenly distribute them in the liquid phase. The purpose of research is thus to find the compounds that provide effective dispersion of carbon nanotubes and simultaneously bind other compounds, including some drugs. Numerous reports concern both covalent or non-covalent modifications of CNT [17].

The covalent modification based on the introduction of functional groups into the graphene surface of carbon nanotubes is an effective way to increase CNT dispersion. This may, however, lead to changes in the physicochemical properties of CNT and create defects in their structure [18–20].

Non-covalent modification is considered a less invasive method. Because of the fact, that there is no heating or acidic environment involved, virtually no damage is done to the structure of nanotubes [8, 21]. Among non-covalent CNT modifications one can distinguish: interactions with amphiphilic molecules – surfactants (e.g. SDS – sodium dodecyl sulphate, SDBS – sodium dodecylbenzenesulphonate, CTAB – cetyltrimethylammonium bromide, DDAB – dimethyldioctadecyl-ammonium bromide, Triton, Pluronic [22]). The most effective interactions between CNTs and surfactants were observed for surfactants containing a benzene ring in the structure (e.g. SDBS), due to the presence of strong pi-pi stacking interactions between the phenyl ring and the graphene nanotube surface [23].

Other compounds used for the noncovalent functionalization and dispersing of CNT include: ionic liquids (ILs) – which form bucky gels, polymers – especially those with aromatic rings in their structure [24], other compounds with an aromatic ring [25] (for example ammonium salts containing pyrene rings [26]) and or deoxycholic acid sodium salts [27].

7.2 The Interaction of Congo Red with Carbon Nanotubes

CR, a bis-azo dye, similarly to the carbon nanotube dispersing compounds mentioned above, has aromatic rings in its structure. It is therefore possible to form a non-covalent pi-pi interaction between CR molecules and the graphene surface of carbon nanotubes. Reports can be found in literature on the interaction of CR with carbon nanotubes [28–31] and the surface of graphene oxide [32, 33].

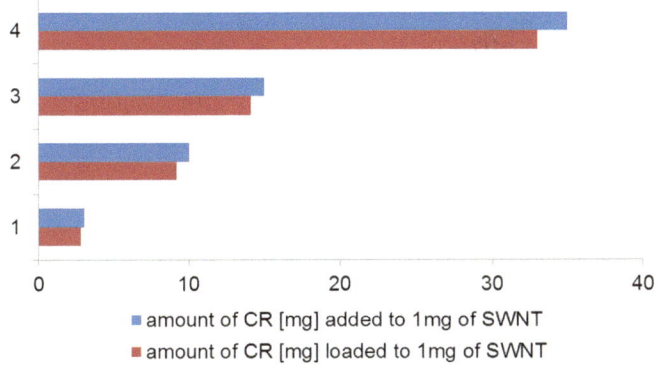

Fig. 7.1 Binding of an increasing amount of CR by 1 mg of SWNT. (1,2,3,4 – samples)

Hu C. et al. [28] describe a complex of single-walled carbon nanotubes (SWNT) with CR (SWNT-CR), with water solubility of up to 3.5 mg/mL. The described complex was formed by prolonged grinding of SWNT and CR in the agate mortar, followed by drying. In this case CR probably interacts with the surface of nanotubes as single molecules bound to the nanotube surface via pi-pi interactions. The described method enables nanotubes dispersion, however the results may not be reproducible and the CR binding effectiveness is relatively low.

In another paper [31] authors point to the possibility of using CNT for adsorbing dyes from aqueous solutions – in post-production wastes containing textile dyes. CR was used as a model compound in these studies and the results show an effective binding of CR by carbon nanotubes (with a maximum adsorption of 500 mg/g).

The SWNT dispersion method that was used in our studies is based on the sonication of SWNT with CR solution and is much simpler and reproducible than the above described method [34]. It leads to the formation of complexes between graphene surfaces of carbon nanotubes and supramolecular ribbon-like structures created by CR molecules. An attempt was made to explain the mechanism of interaction between the ribbon-like supramolecular compounds like CR and the single-walled carbon nanotubes. The results indicate a very efficient binding of CR by SWNT (up to 35 mg CR per 1 mg of SWNT) (Fig. 7.1).

In contrast to the method proposed by Hu et al. the samples are not dried, which allows for efficient binding of CR – in the form of large supramolecular structures – to the SWNT surface. A model of SWNT-CR complex has been proposed based on a "face to face" interaction between CR and the carbon nanotube surface. In this model, the SWNT play a role of a scaffolding to the surface of which CR molecules are bound in the form of supramolecular, ribbon-like structures that protrude from the surface of the carbon nanotube. The nanotube-CR interaction involves the pi-pi stacking between the benzene rings (Fig. 7.2), as in the case of aforementioned surfactants.

The supramolecular character of CR bound to the carbon nanotube surface is indicated by the results obtained during the analysis of SWNT-CR complexes

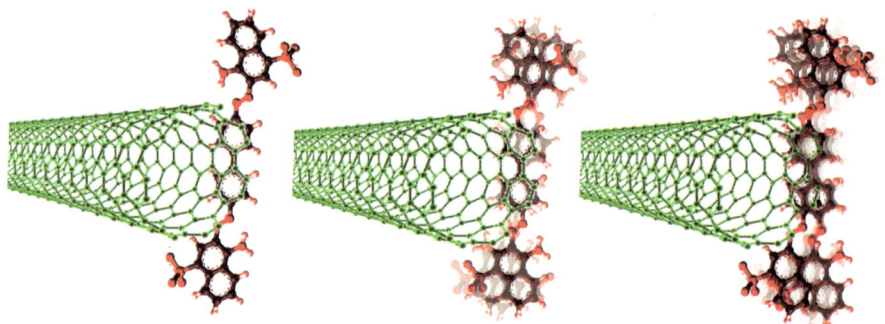

Fig. 7.2 Model of interaction between single CR molecule (*left*) and supramolecular CR (*right*) with the SWNT surface (based on calculations presented in [30])

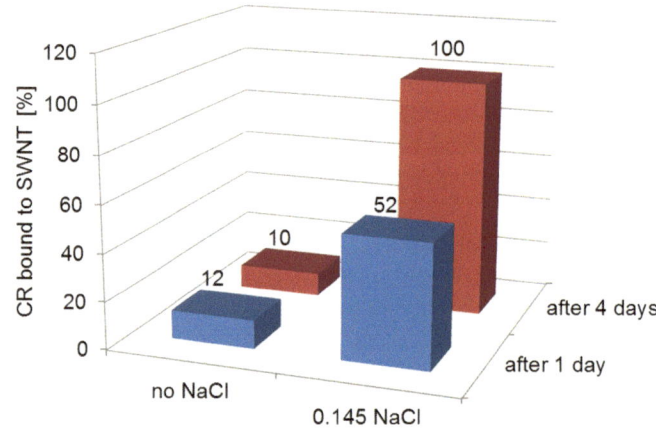

Fig. 7.3 Binding of CR to carbon nanotubes in solutions of different ionic strength. 0.05 M Tris/ HCl buffer pH7.4 without NaCl or with 0.145 M NaCl. The total amount of CR added to the sample was taken as 100%

formed in solutions of various ionic strengths. The properties of CR as a supramolecular system depend on the ionic strength of the solution [35]. If the ionic strength is low, then the ability of CR to form a supramolecular structure is also low, as a result of the lack of shielding of the negatively charged sulphonic groups by Na^+ ions. In contrast, at high salt concentration the supramolecular structure is stabilized. In order to explain if single CR molecules are involved in the SWNT-CR interaction or whether the supramolecular structures are required, an analysis of the SWNT-CR interaction in solutions of various ionic strengths was performed. All samples were prepared in the 0.05 M Tris/HCl pH7.4 buffer. The significant differences in CR binding between samples to which 0.145 M NaCl was added and the NaCl-free samples were observed (Fig. 7.3).

Fig. 7.4 Birefringence of the CR bound on SWNT (polarized light microscopy)

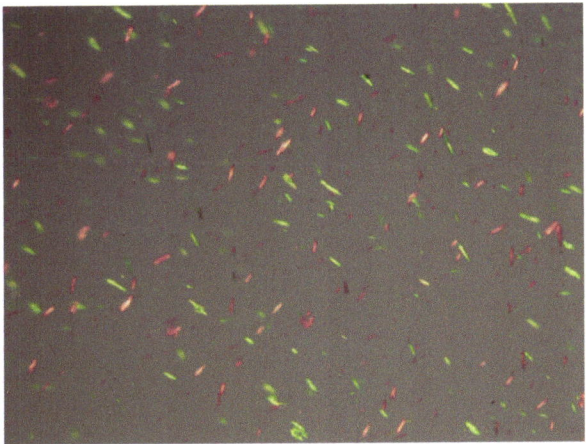

   Additional information concerning the mechanism of CR- nanotube interaction is provided by the birefringence of CR-nanotube complexes observed using polarized light microscopy. This phenomenon, which is typical for the amyloid containing preparations stained with CR, proves that the CR molecules bound to the fibres are ordered [36]. The presence of "apple-green" birefringence in histopathological samples is a diagnostic criterion for amyloidosis. In the case of the SWNT-CR complexes the birefringence also demonstrates the ordering of CR molecules bound to the carbon nanotube surface (Fig. 7.4) and supports the supramolecular, ribbon-like character of CR interacting with the carbon nanotube.

   Samples containing carbon nanotubes sonicated with highly concentrated CR (above 5 mg/mL), presented gel-like properties. CR in these samples was bound to SWNT strongly enough not to be removed during filtration through the PTFEmembrane (non-bonded CR can pass freely through this membrane) (Fig. 7.5A). SWNT-CR complexes in these samples slowly sedimented at the bottom of the test tube but could be easily dispersed even by gentle mixing (Fig. 7.5C). The formation of such big, sedimenting complexes can be explained by the presence of CR ribbons, which protrude from the surface of nanotubes and cross-link them to form stable gels, as in the case of *bucky gels* resulting from the interaction between CNT and ionic liquids (room-temperature ionic liquids, RILs) [24, 37]. In the case of solutions containing a lower concentration of CR the resulting solution is clearer, but also tends to separate over a longer period of time (Fig. 7.5B).

   Spectral analyses have shown changes in the CR spectrum after binding to SWNT, demonstrating the interaction between CR and carbon nanotubes (Fig. 7.6). This effect is related to the change in the electron structure of the CR molecule after binding to carbon nanotubes and is explained by the transition of the CR molecule to the quinoid form [34, 38]. A similar phenomenon is described in the case of binding CR to the surface of graphene oxide [32]. According to the mechanism of the interaction between CR and SWNT described in our paper, not all molecules in the CRribbon-like structure adhere directly to the surface of the nanotube, while the

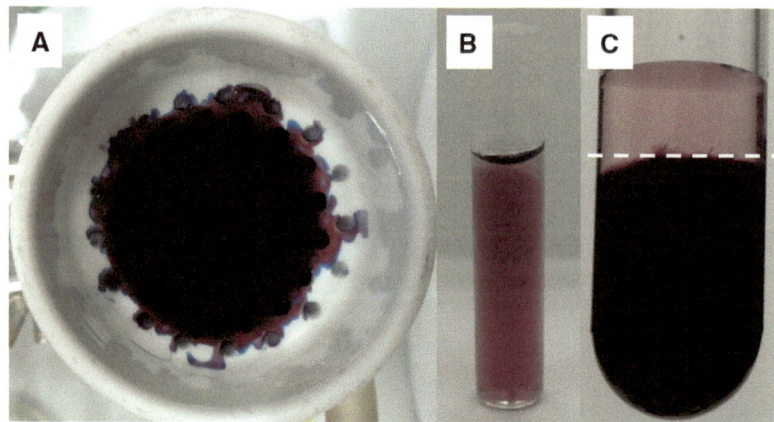

Fig. 7.5 (**A**) SWNT-CR complexes – filtration on PTFEmembrane; (**B**) SWNT-CR complex obtained from 1 mg of SWNT and 2 mg CR, after filtration, suspended in 1 mL of the buffer; (**C**) SWNT-CR complex obtained from 1 mg of SWNT and 5 mg CR, after filtration, suspended in 1 mL of the buffer; the *dashed line* indicates the sedimentation of the gel-like complex

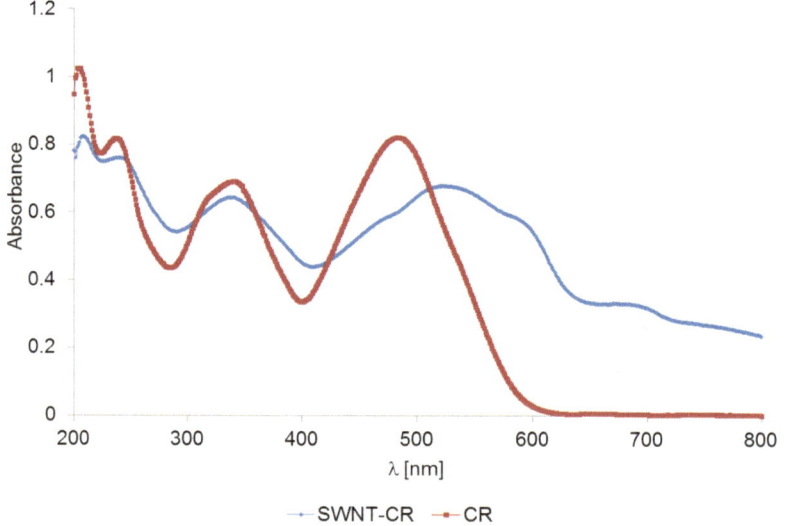

Fig. 7.6 Absorption spectra of free CR and CR complexed with SWNT – maximum of the absorbance shifts from 490 to 530 nm

observed spectral effect – the shift of the absorbance maximum from 490 to 510 nm – suggests that all CR molecules (not only those that directly adhere to the nanotube surface) participate in the transfer of electrons (and switch to the quinoid form) [34].

The SWNT-CR complexes were studied using a variety of microscopic techniques. All analyses have shown that CR binding to the surface of the carbon nano-

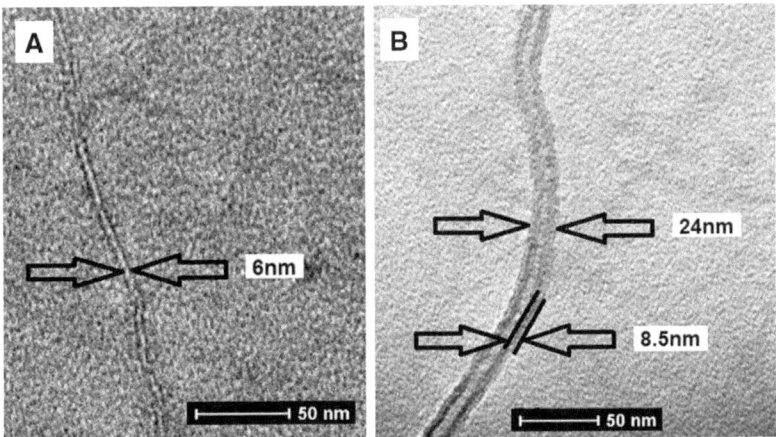

Fig. 7.7 TEM micrographs of (**A**) a single SWNT, without CR; (**B**) a single SWNT functionalized with CR (obtained from 1 mg of SWNT and 10 mg CR, after filtration, suspended in 1 mL of the buffer) – CR layers evenly distributed at the surface of a carbon nanotube are visible (*dark lines* indicate the thickness of CR layer)

tubes significantly increased their diameter. Depending on the amount of CR added per portion of single-walled carbon nanotubes more or less uniformly loaded nanotubes were obtained [34]. The TEM, SEM and AFM images of the SWNT-CR complexes confirm that the interaction between CR and carbon nanotube surface is based on the "face to face" stacking. A comparison of transmission microscopy images of a free carbon nanotube (Fig. 7.7A) and a CR-loaded nanotube (Fig. 7.7B) shows a sixfold increase in nanotube diameter after CR binding. The image shows that addition of highly concentrated CR (10 mg/mL) leads to an uniform covering of the carbon nanotube with CR. This can be interpreted as a large number of short, "face to face" bound CR "ribbons", attached side by side to the carbon nanotube.

The presence of supramolecular CR ribbons that protrude from the surface of carbon nanotubes was confirmed by the results obtained using the ternary complex, composed of CR, Titan yellow (TY) and silver ions (CR/TY/Ag) (see Chap. 6). Titan Yellow-silver ion complex (TY/Ag) intercalates between CR molecules within a supramolecular ribbon. This system was developed because CR molecules do not complex metal ions and thus cannot be directly used for contrasting in TEM images [39, 40]. Thanks to the use of silvercontrast, the binding sites of the ternary complex have been made visible. EDS (energy dispersive spectroscopy) analysis of chemical composition, indicated the presence of silver and sulphur in these sites, which confirms the presence of both CR and TY/Ag. Electronically dense silver ions are visible primarily in places of higher CR/TY loading. Contrasting with silver ions also allowed to confirm the mechanism of supramolecular CR ribbons interaction with carbon nanotubes – silver is not directly attached to the carbon nanotube surface, but is bound to the dye that forms a sheath around the nanotube (Fig. 7.8).

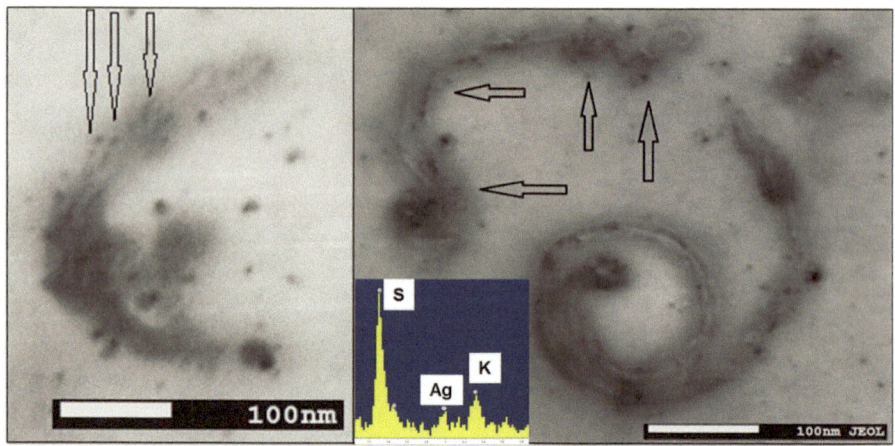

Fig. 7.8 Complexes of SWNT-CR contrasted by intercalated Titan Yellow-silver ion complexes (obtained from 1 mg of SWNT and 5 mg CR/TY/Ag (1:1:0.8 molar ratio), after removal of dyes and Ag excess by filtration). The EDS analysis indicates the presence of CR-TY-Ag complexes in highly loaded places (indicated by *arrows*)

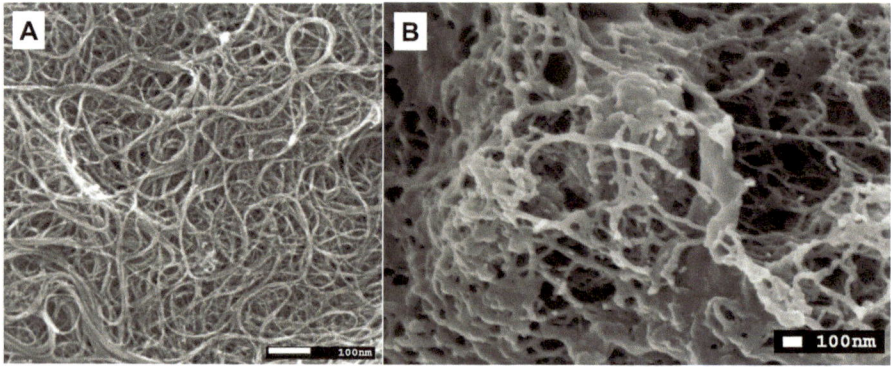

Fig. 7.9 SEM images of (**A**) SWNT and (**B**) SWNT-CR complex (obtained from 1 mg of SWNT and 5 mg CR, after filtration)

The images obtained by scanningelectron microscopy (Fig. 7.9) and atomic force microscopy [34] indicate an increase in carbon nanotube diameter after CR binding to its surface. Unevenly distribution of dye on nanotubes is seen on Fig. 7.9B.

The SWNT-CR complexes were also analysed by molecular modelling, which has shown various possibilities of nanotube-CR interaction depending on nanotube diameter [30]. In the case of narrow nanotubes, CR partially retains the supramolecular ribbon structure, adhering to the outer nanotube surface, while some of the molecules are adsorbed individually on the surface (as shown in Fig. 7.2). In the case of wider nanotubes, individual CR molecules adsorb to the nanotube surface and supramolecular ribbon-like structure of CR is lost. The geometry of the wide nanotube increases the contact surface between CR molecule and the nanotube which allows to ascribe the dominant role in the creation of the system's equilibrium structure to the pi-pi stacking. Nanotubes of smaller diameter have a smaller side wall area that is characterized by a greater curvature. For this reason, partial preservation of the supramolecular ribbon structure of CR is preferable. The molecular modelling results show a tendency for CR to group into ribbons, while simultaneously leaving a portion of the nanotube surface exposed, which was also observed in microscopic images. These results were also confirmed by analysis of the radial distribution functions (rdf) between CR and SWNT [30].

7.3   The Incorporation of Other Compounds by SWNT-CR Complexes – Examples of Possible Biomedical Use

A particularly interesting feature of carbon nanotube-CR complex is its ability to bind other molecules. CNT, as hollow structures, are excellent high volume lightweight containers, in which other molecules can be enclosed while their large surface allows for efficient adsorption of many compounds. CR (and similar compounds) that create ribbon-like supramolecular structures can bind numerous polyaromatic planar molecules that intercalate into supramolecular CR ribbon. Compounds that can be bound this way include e.g. antineoplastic drug doxorubicin, Titan yellow, rhodamine B. The combination SWNT and CR creates a "chemical container" characterized by significantly increased capacity and the ability to bind different molecules simultaneously. It could be used as a drug carrier, which limits the drug's toxicity and allows for the targeted delivery [41].

An interesting phenomenon of breaking carbon nanotubes resulting from the interaction with mixed supramolecular systems was observed. Mixed supramolecular systems (CR/EB and CR-DOX) interact with carbon nanotubes causing them to stiffen and break (Fig. 7.10). Dispersion of nanotubes by CR is not accompanied by their shortening. Similar dye – EB shows much lower nanotube dispersing capabil-

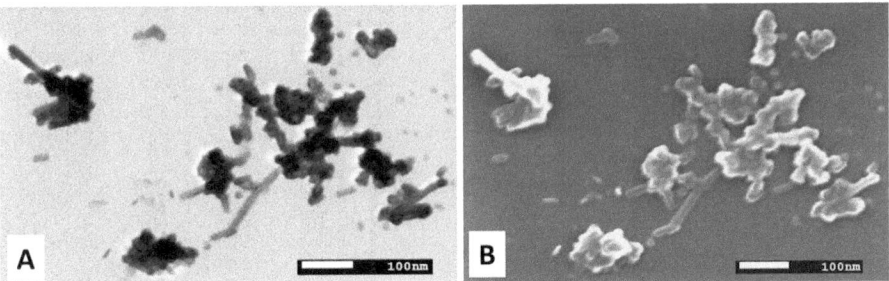

Fig. 7.10 Carbon nanotubes shortened after interaction with the mixed supramolecular complexes of CR/EB

ity and does not interact with their surface [42]. The effect of nanotube breaking can be explained by the mechanical weakening of the nanotube in the area where a large mixed supramolecular system is attached. This phenomenon can also be explained by changes in the electron structure of the nanotube and the supramolecular system bound to it.

The dispersion of carbon nanotubes based on their interaction with CR is simple, efficient and reproducible. Supramolecular systems (both pure CR and mixed) allow not only to disperse carbon nanotubes, but can also bind other compounds through intercalation. As carriers of different compounds (e.g. fluorescent dyes, drugs, metal ions), these complexes present an interesting alternative to the currently used systems. They can be used in diagnostics as well as targeted delivery systems for drugs or metal ions.

Acknowledgements We acknowledge the financial support from the National Science Centre, Poland (grant no. 2016/21/D/NZ1/02763) and from the project Interdisciplinary PhD Studies "Molecular sciences for medicine" (co-financed by the European Social Fund within the Human Capital Operational Programme) and Ministry of Science and Higher Education (grant no. K/DSC/001370).

References

1. Avouris P, Chen Z, Perebeinos V (2007) Carbon-based electronics. Nat Nanotechnol 2(10):605–615
2. Baughman RH, Zakhidov AA, de Heer WA (2002) Carbon nanotubes-the route toward applications. Science 297(5582):787–792
3. Díez-Pascual AM, Ashrafi B, Naffakh M et al (2011) Influence of carbon nanotubes on the thermal, electrical and mechanical properties of poly(ether ether ketone)/glass fiber laminates. Carbon N Y 49(8):2817–2833
4. Guo D-J, Li H-L (2005) High dispersion and electrocatalytic properties of palladium nanoparticles on single-walled carbon nanotubes. J Colloid Interface Sci 286(1):274–279
5. Liu Y, Wang X, Qi K et al (2008) Functionalization of cotton with carbon nanotubes. J Mater Chem 18(29):3454
6. Besteman K, Lee J-O, Wiertz FGM et al (2003) Enzyme-coated carbon nanotubes as single-molecule biosensors. Nano Lett 3(6):727–730
7. Liu Y, Wang M, Zhao F et al (2005) The direct electron transfer of glucose oxidase and glucose biosensor based on carbon nanotubes/chitosan matrix. Biosens Bioelectron 21:984–988
8. Chen RJ, Bangsaruntip S, Drouvalakis KA et al (2003) Noncovalent functionalization of carbon nanotubes for highly specific electronic biosensors. Proc Natl Acad Sci U S A 100(9):4984–4989
9. Ji S, Liu C, Zhang B et al (2010) Carbon nanotubes in cancer diagnosis and therapy. Biochim Biophys Acta Rev Cancer 1806(1):29–35
10. Zanello LP, Zhao B, Hu H et al (2006) Bone cell proliferation on carbon nanotubes. Nano Lett 6(3):562–567
11. Pok S, Vitale F, Eichmann SL et al (2014) Biocompatible carbon nanotube–chitosan scaffold matching the electrical conductivity of the heart. ACS Nano 8(10):9822–9832
12. Aliev AE, Oh J, Kozlov ME et al (2009) Giant-stroke, superelastic carbon nanotube aerogel muscles. Science 323(5921):1575–1578

13. Chen J, Chen S, Zhao X et al (2008) Functionalized single-walled carbon nanotubes as rationally designed vehicles for tumor-targeted drug delivery. J Am Chem Soc 130(49):16778–16785
14. Heister E, Neves V, Tîlmaciu C et al (2009) Triple functionalisation of single-walled carbon nanotubes with doxorubicin, a monoclonal antibody, and a fluorescent marker for targeted cancer therapy. Carbon N Y 47(9):2152–2160
15. Bhirde AA, Patel V, Gavard J et al (2009) Targeted killing of cancer cells in vivo and in vitro with EGF-directed carbon nanotube-based drug delivery. ACS Nano 3(2):307–316
16. Wong BS, Yoong SL, Jagusiak A et al (2013) Carbon nanotubes for delivery of small molecule drugs. Adv Drug Deliv Rev 65(15):1964–2015
17. Pastorin G (2011) Carbon nanotubes: from bench chemistry to promising biomedical applications. Pan Stanford Pub, Singapore
18. Bahr JL, Tour JM, Jakab E et al (2002) Covalent chemistry of single-wall carbon nanotubes. J Mater Chem 12(7):1952–1958
19. Prato M, Kostarelos K, Bianco A (2008) Functionalized carbon nanotubes in drug design and discovery. Acc Chem Res 41(1):60–68
20. Akasaka T, Nagase S, Wudl F (2010) Chemistry of nanocarbons. Wiley, Chichester
21. Backes C, Hirsch A (2010) Noncovalent functionalization of carbon nanotubes in chemistry of nanocarbons. Wiley, Ltd, Chichester, pp 1–48
22. Wang H (2009) Dispersing carbon nanotubes using surfactants. Curr Opin Colloid Interface Sci 14(5):364–371
23. Islam MF, Rojas E, Bergey DM et al (2003) High weight fraction surfactant solubilization of single-wall carbon nanotubes in water. Nano Lett 3(2):269–273
24. Tunckol M, Durand J, Serp P (2012) Carbon nanomaterial–ionic liquid hybrids. Carbon N Y 50(12):4303–4334
25. Tomonari Y, Murakami H, Nakashima N (2006) Solubilization of single-walled carbon nanotubes by using polycyclic aromatic ammonium amphiphiles in water—strategy for the design of high-performance solubilizers. Chem Eur J 12(15):4027–4034
26. Bluemmel P, Setaro A, Popeney CS et al (2010) Dispersion of carbon nanotubes using an azobenzene derivative. Phys Status Solidi 247(11–12):2891–2894
27. Haggenmueller R, Rahatekar SS, Fagan JA et al (2007) Comparison of the quality of aqueous dispersions of single wall carbon nanotubes using surfactants and biomolecules. Langmuir 24(9):5070–5078
28. Hu C, Chen Z, Shen A et al (2006) Water-soluble single-walled carbon nanotubes via noncovalent functionalization by a rigid, planar and conjugated diazo dye. Carbon N Y 44(3):428–434
29. Mishra AK, Arockiadoss T, Ramaprabhu S (2010) Study of removal of azo dye by functionalized multi walled carbon nanotubes. Chem Eng J 162(3):1026–1034
30. Panczyk T, Wolski P, Jagusiak A et al (2014) Molecular dynamics study of Congo red interaction with carbon nanotubes. RSC Adv 4(88):47304–47312
31. Szlachta M, Wójtowicz P (2013) Adsorption of methylene blue and Congo red from aqueous solution by activated carbon and carbon nanotubes. Water Sci Technol 68(10):2240–2248
32. Barkauskas J, Stankevičienė I, Dakševič J et al (2011) Interaction between graphite oxide and Congo red in aqueous media. Carbon N Y 49(15):5373–5381
33. Du Q, Sun J, Li Y et al (2014) Highly enhanced adsorption of congo red onto graphene oxide/chitosan fibers by wet-chemical etching off silica nanoparticles. Chem Eng J 245:99–106
34. Jagusiak A, Piekarska B, Pańczyk T et al (2017) Dispersion of single-wall carbon nanotubes with supramolecular Congo red – properties of the complexes and mechanism of the interaction. Beilstein J Nanotechnol 8(1):636–648
35. Skowronek M, Stopa B, Konieczny L et al (1998) Self-assembly of Congo Red-A theoretical and experimental approach to identify its supramolecular organization in water and salt solutions. Biopolymers 46(5):267–281
36. Spólnik P, Król M, Stopa B et al (2011) Influence of the electric field on supramolecular structure and properties of amyloid-specific reagent Congo red Eur. Biophys J 40(10):1187–1196

37. Fukushima T, Kosaka A, Ishimura Y et al (2003) Molecular ordering of organic molten salts triggered by single-walled carbon nanotubes. Science 300(5628):2072–2074
38. Pigorsch E, Elhaddaoui A, Turrell S (1994) Spectroscopic study of pH and solvent effects on the structure of Congo red and its binding mechanism to amyloid-like proteins. Spectrochim Acta Part A Mol Spectrosc 50(12):2145–2152
39. Chłopaś K, Jagusiak A, Konieczny L et al (2015) The use of Titan yellow dye as a metal ion binding marker for studies on the formation of specific complexes by supramolecular Congo red. Bio-Algorithms Med-Syst 11(1):9–17
40. Rybarska J, Konieczny L, Jagusiak A et al (2017) Silver ions as EM marker of congo red ligation sites in amyloids and amyloid-like aggregates. Acta Biochim Pol 64(1):161–169
41. Stopa B, Piekarska B, Jagusiak A et al (2011) Acta Biochim Pol 58(Suppl. 2):282. Supramolecular Congo red as a potential drug carrier. Properties of Congo red-doxorubicin complexes in Proceedings of the 2nd Congress of Biochemistry and Cell Biology 46th Meeting of the Polish Biochemical Society and 11st Conference of the Polish Cell Biology Society, Kraków, 2011
42. Jagusiak A, Piekarska B, Chłopaś K et al (2016) Shortening and dispersion of single-walled carbon nanotubes upon interaction with mixed supramolecular compounds. Bio-Algorithms Med-Syst 12(3):123–132

Erratum to: Supramolecular Structures as Carrier Systems Enabling the Use of Metal Ions in Antibacterial Therapy

J. Natkaniec, Anna Jagusiak, Joanna Rybarska, Tomasz Gosiewski, Jolanta Kaszuba-Zwoińska, and Małgorzata Bulanda

Erratum to:
Chapter 6 in: I. Roterman, L. Konieczny (eds.), *Self-Assembled Molecules – New Kind of Protein Ligands*, https://doi.org/10.1007/978-3-319-65639-7_6

The original version of the chapter was inadvertently published with incorrect author affiliation. The correct affiliation is as follows:

J. Natkaniec, Małgorzata Bulanda, Tomasz Gosiewski:
Department of Molecular Medical Microbiology, Chair of Microbiology, Jagiellonian University – Medical College, Krakow, Poland

Anna Jagusiak, Joanna Rybarska:
Chair of Medical Biochemistry, Jagiellonian University – Medical College, Kopernika 7, 31-034, Krakow, Poland

Jolanta Kaszuba-Zwoińska:
Department of Pathophysiology, Jagiellonian University – Medical College, Czysta 18, 31-121, Krakow, Poland

---

The updated online version of the original chapter can be found under
https://doi.org/10.1007/978-3-319-65639-7_6

Index

© The Author(s) 2018
I. Roterman, L. Konieczny (eds.), *Self-Assembled Molecules – New Kind
of Protein Ligands*, https://doi.org/10.1007/978-3-319-65639-7